Weather and the Bible

100 Questions and Answers

Donald B. DeYoung

Foreword by Henry M. Morris

BAKER BOOK HOUSE
Grand Rapids, Michigan 49516

a division of Baker Book House Company
P.O. Box 6287, Grand Rapids, MI 49516-6287

Fourth printing, February 1996

Printed in the United States of America

Library of Congress Cataloging-in-Publication Data

DeYoung, Donald B.
Weather and the Bible : 100 questions and answers / Donald B. DeYoung.
p. cm.
ISBN 0-8010-3013-7
1. Bible and meteorology—Miscellanea. I. Title.
BS656.D49 1992
220.8'5516—dc20 92-6248

To

Sally,

my wife and companion

for twenty-six years.

She has been

a faithful partner

through all kinds of weather.

Contents

Part 1 Weather Basics

Part 4 Past Weather

Part 5 Future Weather

Figures and Tables

Figures

Tables

Foreword

This book is a real treasure. It is packed with fascinating information, scientifically accurate, and completely true to Scripture, yet also easy and enjoyable to read. *Weather and the Bible* deals with a subject of universal interest in a uniquely refreshing manner, using a question-and-answer format that is difficult to lay down, yet also nice just for browsing.

Donald DeYoung is a Christian physicist by profession and a Christian gentleman in character. He has already written several books relating science and Scripture, the most recent being *Astronomy and the Bible* (1989), a useful and well-received book of similar format to this present book. He knows science and can communicate it effectively; the same is true of his facility in the Bible.

He has been a respected member of the science faculty at Grace College in Indiana for many years, and is currently editor of the *Creation Research Society Quarterly.* In addition, he is an adjunct professor at the Institute for Creation Research Graduate School of Science in San Diego. Thus Don DeYoung is both an honored colleague and a personal friend.

I am pleased to write a foreword for this fine new book. I commend it heartily to everyone—young and old alike—as a book both to be enjoyed and from which to learn.

Henry M. Morris
President, Institute for Creation Research

Introduction

People talk about the weather more than any other topic, including sports and politics. It is one of the first things we wonder about when a new day begins. To try writing a personal letter without mentioning the weather is a difficult task.

This book was written to help you enjoy some major details of the weather. It is not a professional, comprehensive treatise on meteorology. Such studies often hide the topics that are most interesting. I have attempted to get quickly to the point, whether it is lightning striking particular trees or how Greenland got its name. Literature sources for particular ideas are available upon request. A few selected references appear at the end of the book. Questions with references included are marked "see reference."

Wherever possible, without forcing the issue, I have interacted with the Bible. Scripture provides a rich source for understanding the Creator's provision of the earth's weather system. The reader will notice that the book promotes the recent-creation view of origins. I have found this to be an exciting and insightful way to study nature.

This book has a question-answer format similar to an earlier publication, *Astronomy and the Bible.* Many of the weather questions come from friends in the college classroom and from Bible-science conferences. I have many gracious audiences to thank for encouragement in speaking and writing on the exciting topic of the Bible and science.

The 1990s may be called the "bad-weather decade." We are bombarded with discouraging news regarding global warming, ozone depletion, greenhouse gases, and nuclear winter. Are there any good weather reports in our day? I believe there are, but only in the light of God's Word. The creation, although cursed, still displays its beauty, complexity, and wonder. The weather is no exception; it has thus far confounded our best efforts to control or even accurately predict it. Our precious atmosphere has also shown the strong ability to "bounce back" and adjust to environmental change. Even with one billion exhaust pipes and smokestacks giving off poisons, most locations on earth still have clean air to breathe. How is the weather system designed to give us life? This book explores such questions from the refreshing perspective of creation.

Acknowledgments

My special thanks to five meteorologist friends who provided helpful suggestions: Charles Clough, Craig James, David Newquist, Michael Oard, and Larry Vardiman. Editorial help was also provided by John Davis, Ivan French, David Rodabaugh, and John Whitcomb. Of course, all errors and personal viewpoints are my own. Thanks to Virginia Larson for typing several drafts of this manuscript.

PART **1**

Weather Basics

1. Should a Christian become a meteorologist?

Many Christians have become suspicious of science, and rightly so. The rise of technology has paralleled a decline in respect for Scripture. Scientific discoveries are frequent in our day, but thanks is seldom expressed to the God who makes these discoveries possible in the first place. Space probes explore the universe, but man gets all the glory rather than God. For these reasons, if no other, Christians are indeed needed in science careers today. In laboratories everywhere is a desperate need for the perspective that God designed this universe, that he owns it, and that we are accountable stewards.

When God gives a young person special talent and interest in weather studies or any other technical field, it should certainly be pursued as a basis for a possible career. The study of nature is highly recommended by Scripture:

> The heavens declare the glory of God;
> the skies proclaim the work of his hands (Ps. 19:1).

> Go to the ant, you sluggard;
> consider its ways and be wise (Prov. 6:6).

> It is the glory of God to conceal a matter;
> to search out a matter is the glory of kings
> (Prov. 25:2).

> See how the lilies of the field grow. They do not labor or spin. Yet I tell you that not even Solomon in all his splendor was dressed like one of these (Matt. 6:28).

The Christian scientist is in good company. Many of the greatest scientists of all time gave testimony of personally knowing the Creator: Johann Kepler, Isaac Newton, James Clerk Maxwell, Michael Faraday, and others.

Some present-day skeptics claim that the Christian cannot really do science because his mind is already made up concerning such things as origins and the truth of the Bible. Such skeptics have apparently forgotten that the entire foundation of modern science was laid down by devout Christians. These modern-day skeptics have also failed to look inward to recognize their own particular presuppositions. No scientist enters the laboratory completely unbiased to all ideas. His or her particular world view will be reflected in the choice of experiment and also in the interpretation of results.

In past centuries, weather forecasting was viewed with suspicion. The practice of predicting future weather seemed to border on witchcraft. However, meteorology (study of weather) has today become a well-established application of science. Weather forecasting is complicated and inexact, but it is a very useful and honorable profession. A career in meteorology could be an excellent ministry. Many radio and television weather reporters indeed have a valuable Christian testimony.

A new generation of science workers is needed who will bring honor to the Creator. Ministry opportunities in science are unlimited: worldwide interaction with students, faculty, and other researchers; researching the details of the creation; sharing these truths with others.

2. Is there a creationist view of the weather?

The creationist view of the weather, and of science in general, has several distinctives. These distinctives do not hinder the pursuit of truth. Instead, they provide a refreshing approach to scientific data. The distinctives include at least the following:

The Bible is accepted as a literal, true record of history. The physical universe, including the earth and its atmosphere, was supernaturally made in just six days from nothing. The Creator continually maintains the constancy and integrity of his natural laws, energy, and matter. He has also intervened in earth history on many occasions, including miracles and supernatural weather modification: the Genesis Flood, plagues of the Exodus, fire from heaven upon Sodom and Gomorrah, calming of the sea. The secular alternative is a self-generating universe with no outside intervention, a "closed" universe that is entirely independent of God.

Creationists see intelligent design in nature. Planning and purpose are in all things. The master designer arranged the complex materials and weather systems of the earth, and this makes scientific study a worthy endeavor. In contrast, secular science too often advocates an accidental, chance, random, spontaneous development of the earth and universe. Seeing no plan or purpose, yet faced with intricate natural laws, many scientists can do no better than suggest some mystical, unknowable force within nature.

Instead of slow changes over billions of years, many creationists accept a fully functioning, mature beginning for the earth and space. In this view, the earth's history is only thousands of years, not multiple billions. Earth structures such as sedimentary layers and folded rock strata are due to major catastrophes, primarily the Genesis Flood. The collapse of a vapor canopy (Q. 74) may have led to significant changes. In the creationists' view, the past is the key to understanding the present earth. This is a reversal of geologists' common premise that the present is the key to the past.

The original creation was perfect and "very good" in every way. The world was soon marred by the curse upon nature, a result of the sin of our first parents. However, the weather and creation in general continue to reflect integrity and goodness. All parts of the creation work together, like the gears of a complex machine. When we change one part of the system in an effort to "improve" things, negative consequences always arise, often in unexpected places.

The creationist viewpoint sees humanity as the culmination of the creation week. Planet Earth was made for our enjoyment and well-being. The water cycle and the cool breezes were planned for our benefit. The trees were made for our visual enjoyment, as well as for food (Gen. 2:9). In contrast, a consistent evolutionist view makes nature entirely independent of humanity: If we had not evolved, history would still have continued on earth in other directions.

Many further distinctives could be listed, but these five—supernatural beginning, design, youthfulness, perfection, humanity centered—will suffice to show that the creationist approach is well-defined and consistent.

3. How do climate and weather differ?

Winters often seem to be unusual. They are very mild or else they are extra severe. When this happens, thoughts quickly turn to permanent, long-term changes in the weather: "It used to be colder, or warmer, or dryer, or wetter." Actually, the weather for one particular season is a poor indicator of permanent change. *Weather* describes what is happening in the atmosphere today. *Climate* describes a long-time average of the weather, usually covering thirty to one hundred years. Daily weather continually fluctuates around the climate average. A cold winter does not mean that a hot summer will immediately follow. Over the years, however, extra cold and extra warm winters will tend to cancel each other out and be part of the average.

Weather compares with climate as the experiences of one day compare with a lifetime. A particular day may be normal or may be filled with special events such as parties and presents. Over a lifetime, the daily emotional highs and lows blend together to form an overall outlook on life. Likewise, climate is an overall description of a region's weather.

The definition of climate as a weather average raises two cautions regarding present environmental concerns. First, dire predictions of global warming or even an impending ice age are very uncertain. Until many years of weather data are collected, trends are simply not known. Second, if we do indeed eventually change the climate, perhaps through an enhanced greenhouse effect (Q. 93), the change will not easily be undone. Climate shifts are gradual, and future generations may have to live with whatever climate they inherit from us.

4. Why are weather predictions uncertain?

Meteorologists have learned to forecast weather accurately a few days in advance. Longer than this, however, weather prediction becomes very uncertain. Two reasons account for the failure of long-range accuracy. First, weather is the result of a great number of variables or controlling factors. Most basic are temperature, air pressure, moisture, and wind. Also involved are the jet streams, ocean currents, and land elevation. As the number of such variables increases in any scientific endeavor, the results become less predictable. These multiple variables rule out any comprehensive mathematical formula by which next month's weather can be exactly calculated. Instead, computer models are used, with varying levels of accuracy. The effort to predict weather was one of the initial reasons for the development of computers.

Second, weather details are ruled by what science calls unpredictable "chaos." This means that very small environmental factors can eventually result in large-scale differences in weather patterns. Future weather is sensitive to initial conditions, which can never be known exactly. No matter how good the initial data, there is an inherent limit on an accurate prediction. As an extreme example, somewhat overstated, the flapping of a butterfly's wings in the Amazon rain forest may set in motion a chain of events that later results in a Midwest tornado—appropriately called the butterfly effect but more grandly known as the law of unintended consequences. Since infinitesimal beginnings can later have magnified results, precise weather forecasting beyond a few days will remain impossible. There is no way to determine exactly what the weather will be a year from now. Only the Creator knows future events exactly, including the weather. He allows the flutter of the butterfly's wings.

What about the *Farmer's Almanac,* where weather for the entire year is predicted? If it appears to be accurate, this is only because the forecasts are written in a general way: partly cloudy, chance of rain, etc. Almanacs are based on average climate data, but long-range, day-by-day weather descriptions are only guesses.

5. Are traditional weather proverbs accurate?

Tradition has a long list of weather forecasters, ranging from woolly bear caterpillars to aching teeth. Some come from long experience and indeed have factual bases. Three examples will be considered here.

First, the Lord Jesus mentioned well-known signs in the sky to his critics:

> The Pharisees and Sadducees came to Jesus and tested him by asking him to show them a sign from heaven. He replied, "When evening comes, you say, 'It will be fair weather, for the sky is red,' and in the morning, 'Today it will be stormy, for the sky is red and overcast.' You know how to interpret the appearance of the sky, but you cannot interpret the signs of the times" (Matt. 16:1–3).

This particular weather prediction is more commonly stated:

> Red sky at night, sailor's delight;
> Red sky at morning, sailors take warning.

A red sunset results when the western sky is especially clear. The color occurs because the sun is low in the sky and its light passes through additional atmosphere (see Q. 21). The condition is enhanced if a stable high pressure region is present. This high pressure suppresses cloud formation and also holds air contaminants near the earth. These in turn "scatter" the colors of sunlight and cause the reddening effect in the west. In the mid-latitudes of the northern hemisphere weather patterns usually approach from the west. Since "highs" generally bring good weather, red skies in the evening indicate that fair weather probably is approaching from the west. On the other hand, if the red appears in the eastern morning sky, then the high pressure region has already passed through. A low pressure system generally follows a high, and this low pressure is often associated with unsettled, stormy weather. Sometimes, evening redness is also due to sunlight reflecting off a retreating cloud layer in the east. Likewise, morning redness in the west may be due to an advancing cloud layer. Note the Lord's application of this common weather prediction: His critics readily accepted these natural signs in the sky, but they did not accept him or recognize the important spiritual events of their own day.

A second example of traditional weather prediction is to go by the aches and pains that many people feel before a storm arrives. This is a genuine effect associated with the barometric pressure. Our bodies have "built-in barometers." As mentioned before, low pressure usually precedes a storm. When the barometer falls, tiny bubbles of gas within the body may expand and exert a painful force against nerves. This is especially noticeable around a bunion, rheumatic bone, joint, or sensitive tooth. Weather changes also give some people headaches, but this affliction is not well understood.

A third example of traditional weather prediction is various animals' behavior in nature. Groundhogs, caterpillars, ducks, and fiddler crabs have all been studied for their weather secrets. Air pressure sometimes affects animal behavior:

Swallows fly high, clear blue sky;
Swallows fly low, rain we shall know.

Many animals have very acute senses. Fish, for example, respond to small pressure changes; sometimes they bite, at other times they don't. The activities of birds and animals are occasionally used to predict the severity of entire seasons, but the validity of these long-range predictions is questionable. If a squirrel gathers extra acorns, it does not necessarily mean a severe winter is approaching. Instead, there may simply be a bumper crop of acorns. If animals have instinctive insight into long-range weather patterns, we are not yet familiar with their methods. In studying animal behavior and weather, it is sometimes difficult to separate folklore from facts.

Groundhog Day, February 2, is about halfway between the start of winter (December 21) and spring (March 21). It actually began as a religious holiday called Candlemas. One traditional explanation is that it commemorated the presentation of the infant Jesus at the temple, when Simeon declared that Jesus would be a light for revelation to the Gentiles (Luke 2:32). The early church celebrated this event with an annual parade of candles. In time, the day assumed prophetic status for the months ahead. It is uncertain how the groundhog took over the day. However, his shadow means no more than your own.

In general, many traditional weather predictions are quite accurate. Some, however, have no factual basis.

6. Do moon phases control our weather?

Phases of the moon are half-seriously blamed for a large variety of daily events. Stock market changes, birth rates, mood changes—all have been connected with the moon's appearance in the sky. Actually, there are only two known physical relationships between the earth and the moon: the night light and the tides. Most stories about lunar influence are rather subjective. For example, during one particular month the full moon may coincide with many hospital births, and everyone remarks about the moon's effect. Next month the full moon may occur with normal or fewer births, and no one notices.

A connection between the moon and the earth's weather must similarly be placed in the doubtful category. Statistical studies have shown no definite correlation between moon phases and our weather. Granted, there is much we don't know about the moon. A weather connection with the moon is actually an old idea. This can be seen in the centuries-old names given to the lunar plains that cover the visible surface of the moon. These plains of hardened lava were originally called seas by Galileo, from their dark appearance in early telescopes. The first- and third-quarter moon phases, which occur two weeks apart, show surface features with the following names (figure 1):

First Quarter	*Third Quarter*
Sea of Tranquility	Sea of Clouds
Sea of Nectar	Sea of Storms
Sea of Serenity	Sea of Cold
Sea of Fertility	Sea of Showers

Obviously, the traditional view related good weather with the first-quarter phase, and a time of storms with the third quarter moon. An interesting project would be to monitor calendar moon phases over a period of time and record the accompanying weather. Past studies show no definite weather correlations with moon phases. The same moon phase occurs everywhere on earth, and there is always great variety in local weather conditions.

On certain occasions the moon is able to intensify stormy coastal weather damage. Each month there are two especially high tides, occurring during the new and full moon phases. If a storm happens to accompany these maximum tides, waves may be exceptionally destructive to shoreline property. Tides also occur in the earth's atmosphere. The air rises and lowers, somewhat as do the ocean tides. Atmospheric tides possibly have some effect on the jet stream (Q. 38), or on high and low pressure regions.

Figure 1

The pictured moon phases occur about 3-1/2 days apart.

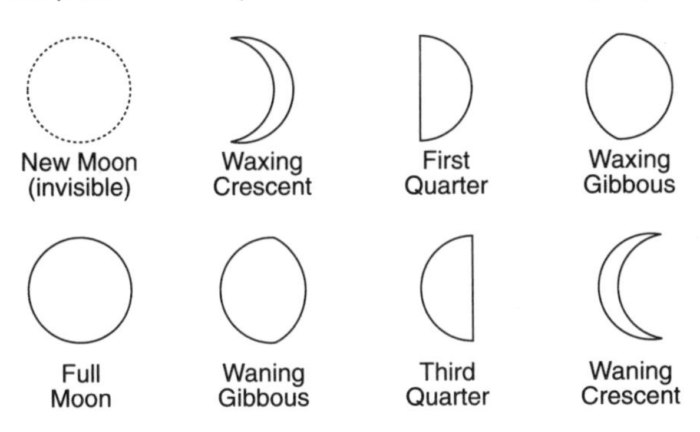

7. What is the atmosphere?

The atmosphere is a thin blanket of gas that surrounds the earth. Ninety-nine percent of our air is within 25 miles of the earth's surface. If a two-inch circle is drawn with a sharp pencil to represent the world, then the depth of the atmosphere will be less than the thickness of the curving pencil line. Our atmosphere is precious and also quite limited in extent.

The atmosphere is divided into several layers (figure 2). The lower layer, about the first 6 miles, is called the troposphere. This is the region where most of our weather occurs including winds, clouds, and rain. It is also the highway for most airplane traffic. The stratosphere extends about 6–30 miles high. This name describes flatness, because the air in this region is visibly layered. The record manned balloon ride was 21 miles high, extending well into the stratosphere. This flight was made in 1961 by a naval officer who brought along warm clothes and an oxygen supply. If he had skydived to the ground from the stratosphere, the trip would have taken five minutes. The stratosphere region is also home to the important ozone layer. Next above the stratosphere are the mesophere and ionosphere. The air at these altitudes is very thin, or rarefied. Some of these upper air molecules escape completely into space. The northern lights, or aurora, also occur in these upper air layers, 60–600 miles high. These lights are caused by charged particles from the sun that collide with earth's air molecules and make them glow. See Q. 8 on the actual composition of the atmosphere.

The apostle Paul refers to the extent of the atmosphere in 2 Corinthians 12:2. He describes a man being "caught up" to the third heaven. This may have been Paul's view of the three heavens:

First heaven. The entire atmosphere as we know it today. It includes the air and clouds, wind and weather.

Second heaven. The vast region of space including the sun, moon, and stars.

Third heaven. The throne of God, beyond the visible realm. The actual location is unknown; Scripture tells that our Lord ascended far above all heavens (Eph. 4:10).

The apostle Paul's description of the three heavens is entirely consistent with our present view of the vast creation.

Figure 2

The division of the atmosphere into layers. Our weather takes place in the lower layer, the troposhere. Note that the vertical scale is logarithmic. Mt. Everest is about 5.5 miles high.

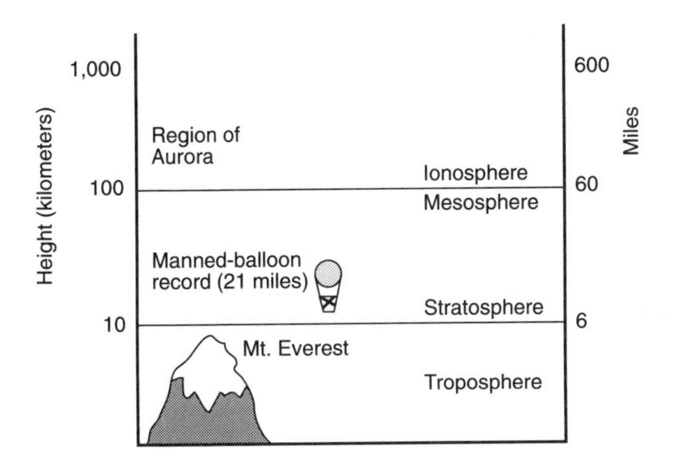

8. What is the composition of air?

The earth's atmosphere is a mixture of several different gases. Clean, dry air has the following composition by volume (figure 3): nitrogen, 78 percent; oxygen, 21 percent; argon, .93 percent; carbon dioxide, .035 percent; other gases, .035 percent.

From the perspective of creation this distribution of gases is not accidental but shows planning and design. People and animals utilize the oxygen portion of air. The abundant nitrogen is simply breathed in and out with no effect. However, this is as it should be. If oxygen were the dominant gas in the atmosphere, spontaneous fires would break out worldwide. Oxygen is very flammable, while nitrogen is not. It thus appears that nitrogen has been provided to safely dilute the oxygen. Calculations show that an atmosphere with 25 percent oxygen might be explosive. At the other extreme, a reduction to 15 percent oxygen would be suffocating for all air-breathing life on earth. Hence the 21 percent figure for oxygen is an optimum value, planned for our well being. A similar balance applies to carbon dioxide in the atmosphere, the gas "breathed" by vegetation. If the carbon dioxide amount in the air were larger, the earth might become warmer (Q. 93). If the carbon dioxide amount were lessened, plants would starve. The gas called argon can be separated from the atmosphere with many practical uses—for example, in the welding of certain metals. The actual role of argon in the atmosphere is not well understood, but it is probably beneficial to life on earth in many indirect ways.

The "other" category of atmospheric gases includes water vapor, neon, helium, methane, krypton, hydrogen, carbon monoxide, xenon, ozone, and radon. The tiny amount of atmospheric water has an awesome effect on our lives. Weather is largely due to this meager compo-

Figure 3

The distribution of gases in the earth's atmosphere by volume. Nitrogen dominates the air, followed by oxygen.

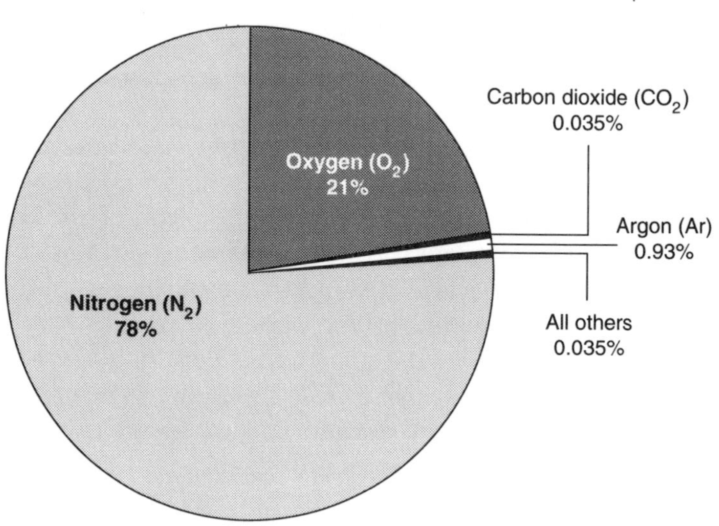

nent of the air. It would be interesting to explore the importance of all the other minor gases to our well-being. On the created earth, no details have been left to chance. (See Q. 89 for a discussion of the atmosphere in the past.)

9. Does air have weight?

It may seem that the invisible air is entirely weightless. However, each cubic foot of air actually measures about one-tenth pound. The air in a typical room adds up to several hundred pounds, and the entire atmosphere amounts to five billion million tons. That is about a million tons of air for every person on earth. Since a person breathes only a few hundred tons of air during a lifetime, and even this amount is recycled, there is plenty of air for all of us. The earth's blanket of air results in a surface pressure of about 15 pounds to the square inch. We walk around at the bottom of this "ocean" of air without noticing the pressure.

Barometers measure this air pressure and record its slight changes as weather patterns come and go. Air pressure affects many things, and may even determine whether or not your homemade cake rises. Early barometers were called weather glasses. The air pressure pushed liquid from a reservoir up into a glass spout. Today, most home barometers have a small sealed box that expands and contracts with pressure changes. An attached needle then registers the air pressure reading. The weather reporter gives air pressure to the nearest one-hundredth inch of mercury, such as "30.12 inches." This is the depth of mercury which exerts the same pressure as the air.

High pressure results when more dense air is present. It will be replaced, sooner or later, by a low pressure region. The change in barometric pressure is especially important in weather forecasting. When pressure is rising, clear skies probably lie ahead. High pressure systems are often associated with sinking air, which inhibits cloud formation. The high pressure may also impede stormy weather from moving in. A decreasing barometer is a warning of rough weather ahead. Low pressure areas usually contain warm, rising air. When this air reaches saturation, it forms clouds and precipitation.

A reference to the weight of the air is found in Job 28:25, 27: "When [God] established the force of the wind and measured out the waters . . . then he looked at wisdom and appraised it."

The force of the wind (*weight* in the King James Version) accurately describes the heaviness of the air. If air had no weight, then wind could have no force. Job 28 is entirely correct in referring to the force or weight of the wind. In this technical age it is often said that the Bible is out-of-date and "pre-scientific" in its views. After all, Scripture was written many centuries ago, before modern discoveries. In truth, however, God's Word is found to be entirely accurate, even in the smallest details such as the weight of the air.

10. How much sunlight hits the earth?

Very little of the sun's total energy actually hits the earth. Figure 4 shows the small target the earth makes as solar energy spreads outward in all directions. We can calculate the fraction of total sunlight energy that continually hits the earth (r = earth radius, R = earth-sun distance):

$$\frac{\text{Cross section area of earth}}{\text{Area of sphere with radius R}} = \frac{\pi r^2}{4 \pi R^2}$$

$$= \frac{\pi\,(3960 \text{ miles})^2}{4 \pi\,(93{,}000{,}000 \text{ miles})^2}$$

$$= .45 \times 10^{-9}$$

The result shows that slightly less than one-half billionth of the sun's energy actually contacts the earth. A critic of creationism once told me that this process proves the Creator is wasteful of energy. However, the Creator has infinite energy resources at his command, which leaves the criticism without meaning. Furthermore, we couldn't handle any additional solar energy. If the earth absorbed just one additional billionth of the sun's total energy, earth temperatures would soar and our planet would bake.

Of the sunlight that does hit the earth, only half is actually absorbed at the surface. The rest is directly reflected back into space by the clouds, air, and ground. This dilution of solar energy helps maintain comfortable temperatures on earth. On an annual average, solar power hitting the earth is about 30 watts to the square foot. If all

Figure 4

Our planet is a very small target for solar energy. The earth is actually 109 times smaller than the sun. If the sun were basketball size, then the earth would only be one-tenth inch and would be positioned one hundred feet from the basketball.

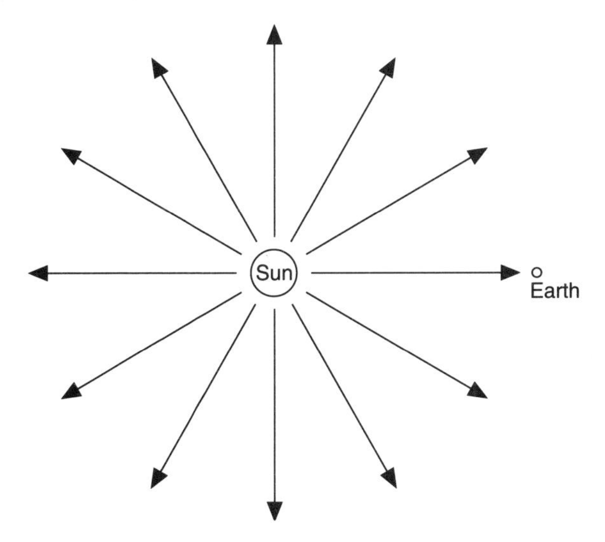

this power could be captured with solar cells, a football-size area could supply energy to a small city. In future years, use of solar energy for electricity may expand greatly. This is the main source of energy God provides for us.

11. What is temperature?

Temperature is not a simple concept to define. It is actually a measure of the amount of motion, or kinetic energy, of molecules. If an object is warm, its molecules move more vigorously; they have more energy. When molecules increase their motion they also tend to spread out. This is called "thermal expansion" and is the basis of most thermometers. You have noticed how mercury or some other fluid expands and climbs up the thermometer tube on warm days.

Several different temperature scales have been defined by calibrating thermometers with numbers. Two useful constant temperatures are the freezing point of water (32°F; 0°C) and the boiling point (212°F; 100°C). The two popular temperature scales, Fahrenheit (T_F) and Celsius or Centigrade (T_C), are related by these equations:

$$T_F = 9/5\,T_C + 32$$

$$T_C = 5/9\,(T_F - 32)$$

At just one particular cold temperature, both of these scales give the exact same reading, −40°F = −40°C. Above this point, the Fahrenheit temperature is always greater than the Centigrade value.

The following list gives a few temperatures of special interest:

11,000°F; 6000°C Surface of the sun. The solar interior is millions of degrees.

95°F; 35°C The Red Sea, warmest sea on earth. Many hot-water springs feed into it.

72°F; 22°C Ideal indoor temperature.

40°F; 4.5°C Ideal temperature for storing milk and food.

39°F; 4°C The temperature where water is most dense. Colder water expands, thus causing ice to float (Q. 32).

30°F; –1°C Arctic water, sub-freezing, but not ice because of its salt content. Caused unconsciousness within minutes for the victims of the *Titanic* sinking.

–22°F; –30°C Daytime temperature on Mars, measured by the *Viking* probes.

–108°F; –78°C Frozen carbon dioxide, also called dry ice.

–306°F; –188°C Liquefied oxygen.

–459°F; –273°C Absolute zero; the limit of low temperature. All atom vibrations cease.

Just as we are sensitive to different temperatures, animals also notice heat changes. This results in some unusual "thermometers," including the use of crickets:

$$\text{Temperature in °F} = \text{number of cricket chirps in 14 seconds} + 40$$

If a cricket chirps 20 times in 14 seconds, the temperature should be close to 60°F. Do the crickets in your neighborhood obey the rule? Ants also provide a thermometer; they crawl faster as the temperature increases. However, you will have to make up your own formula for ants.

12. What is a temperature inversion?

When the residents of Donora, Pennsylvania, awoke on October 26, 1948, they noticed an unpleasant smell. This was nothing new since the city is located in the Monongahela Valley and is surrounded by smokestacks from the steel plants. This morning, however, sulfur dioxide fumes thickened. A severe temperature or thermal inversion had trapped the polluted air in the valley. The birds were silenced. For the next five days, a fog of unhealthy chemicals accumulated. Of Donora's 14,000 residents, half became ill. About twenty people who were especially sensitive to the unclean air died, mainly the elderly.

Usually, the temperature of air decreases as it moves upward in the atmosphere. The amount of change is called the normal lapse rate—about 5.5°F per 1000 feet for dry air and 3.5°F for moist air. Thus, if you climb 10,000 feet, the temperature is about 55°F colder. Sometimes, however, cool surface air is trapped beneath a lid of warmer air above. It is called a temperature inversion, because it is inverted from the usual condition. Vertical mixing and horizontal winds are typically weak under such conditions. Exhaust chemicals injected into this surface air are unable to disperse and dilute, often leading to a severe pollution episode as in Donora.

A thermal inversion can develop in several ways. A surface inversion may occur near the ground on clear, calm nights. In the evening heat radiates from the ground and its surface cools more rapidly than the upper air. When the surface again warms the next day, the inversion usually disappears. In valleys such as Monongahela, however, it may linger and reduce air quality by a lack of circulation. An aloft inversion is caused by sinking air from a high-pressure center. As the mass of air descends it is compressed and warmed. An inversion develops between the subsiding warm layers and cooler air trapped beneath.

This second type of inversion may occur over a large region and last for several days.

Freezing rain in wintertime may also be due to a temperature inversion. Raindrops first form in the warmer air above. When they encounter the colder ground temperatures they quickly freeze.

13. What are some weather extremes on earth?

The earth has a vast surface area, nearly 200 million square miles. A great variation in weather covers this expanse. The following list describes just a few places with unusual distinction:

Northern Chile is one of the driest places on earth. Cold ocean currents prevent the formation of clouds, and in places there is no recorded rain in over 400 years. Ground water and dew formation allow the growth of hardy grasses.

Central Uganda in Africa averages 242 thunderstorm days each year.

Mt. Waialeale in Kauai, Hawaii, averages 460 inches of rain (38 feet) each year. At an altitude of nearly a mile, the cold temperatures squeeze the moisture from rising air.

The highest air temperature ever recorded was 136°F, in Libya. This occurred in 1922, in the shade. Death Valley, California, has recorded a temperature of 134°F. The lowest reading ever measured was -129°F in Antarctica, in 1983.

In 1921, 6.3 feet of snow fell at Silver Lake, Colorado, in 24 hours.

Holt, Missouri, experienced 12 inches of rain in just 42 minutes, in 1947.

People live near each of the above locations, in spite of the weather extremes. The Creator has also given an abundance of plants and animals that thrive in seemingly impossible places. In comparison with the harsh conditions of other planets (Q. 28), no place on earth is totally desolate or lifeless.

14. What causes dew and frost?

Dew and frost result from the condensing or freezing of water that is present in air. This happens when the atmosphere becomes saturated, that is, when the relative humidity becomes 100 percent. Relative humidity is a measure of how much moisture is in the air, compared with the maximum amount of moisture the air can possibly hold at that temperature. Relative humidity is low in the desert, and high on a humid summer day in the Midwest.

Dew forms when the ground surface becomes cool during the night hours and condenses the moisture from the air that touches it. This occurs often in late summer and fall while the atmosphere is still warm and filled with moisture. In the morning when the ground warms, the moisture evaporates once again. Perhaps you have seen dew droplets on grass, spider webs, and park benches. You may also have noticed "dew" form on a bathroom mirror. Warm, moist air touches the cool mirror, and water droplets form.

An interesting reference to dew is in Judges 6:36-40. Gideon asked a sign of the Lord that Israel would be victorious in battle. Gideon requested that morning dew be formed on a piece of wool but not on the nearby ground, and it was so. For the next morning Gideon asked the reverse, dry wool and wet ground, and it also happened. As promised, the enemy was indeed delivered into the hands of the Israelites. This sign was a miracle, since dew would not naturally form on alternate surfaces.

Frost represents a special cold temperature change called deposition. In this case, the gaseous water vapor converts directly to its frozen, solid form without passing through the intermediate liquid state. Frost is a coating of very small ice crystals, sometimes with intricate snowflake patterns. It is sometimes called hoarfrost when the ice crystals grow especially large and white. Psalm 147:16 describes God's beautiful painting of the landscape with ice: "He spreads the snow like wool and scatters the [hoar] frost like ashes."

15. Does springtime drive the frost deeper?

Buried water pipes sometimes survive the winter, only to freeze and crack in the spring. Thus it seems that the cold is pushed farther into the earth toward the end of winter. Actually, however, cold temperatures gradually work downward throughout the winter season. The ground is a good insulator, and freezing temperatures may not reach buried pipes until spring, when the coldest weather is already past.

The surface of the earth experiences a similar delay in temperature change. In the northern hemisphere our longest period of daylight occurs on the first day of summer, about June 21. After this "summer solstice," the northern hemisphere begins to tilt away from the sun (Q. 16), and days shorten. However, our warmest days usually occur weeks later, during July and August. This time delay results from the gradual warming and cooling of this part of the earth. The land and water slowly absorb heat, and they later give this heat back to the air. This process ensures that seasonal changes are slow and dependable.

16. Is the earth closer to the sun during summer?

At a recent commencement at Harvard University, this question was asked of twenty-three gowned graduates. Twenty-one answered incorrectly, saying that the earth is closer to the sun in the summer. Actually, the earth is 4 million miles farther from the sun during the summer season of the northern hemisphere. During an entire year the earth-sun distance varies between 91 and 95 million miles, a 4 percent change. It is the tilt of the earth's north-south axis that controls our seasons rather than the small distance change.

Figure 5 shows the tilt of the earth's north-south axis, 23.5° with respect to its orbit plane around the sun. Because of this planetary tilt, the sun appears to gradually move back and forth across the equator during the year. In the northern hemisphere, summer occurs while that part of the earth tilts toward the sun. Even though the earth is slightly farther from the sun during this season, the sun rises higher in the sky in summer and warms the land.

If the earth's tilt were greater than 23.5°, our seasons would be more severe. Consider the planet Uranus with an axis tilted completely sideways, 98°. This causes its north and south poles to alternately point directly toward or away from the sun as it "rolls" around its orbit. On the earth such a severe tilt would cause large-scale melting and refreezing of the polar ice caps. The continents would be periodically flooded, and would also experience long periods of darkness. Actually, since Uranus is nineteen times farther from the sun than the earth, the only seasonal change there is from cold to colder. In contrast to Uranus, Venus is a planet with no axis tilt at all. The sun remains above the equator continually with no seasonal changes in its northern and southern hemispheres. If this were true on earth, the equatorial areas would grow hotter and the ice caps would probably expand. Much fresh water would be permanently frozen, leading to expanded deserts and smaller seas. It is difficult to imagine the struggle for life on the earth if its axis tilt were changed in either direction, smaller or larger. Such details as the earth's tilt were surely designed with our welfare in mind.

Figure 5

As the earth circles the sun, its north-south axis maintains a 23.5° tilt. The earth is slightly farther from the sun during the northern hemisphere's summer. The earth is shown at the four positions that start the seasons on the approximate dates of June 21, September 21, December 21, and March 21.

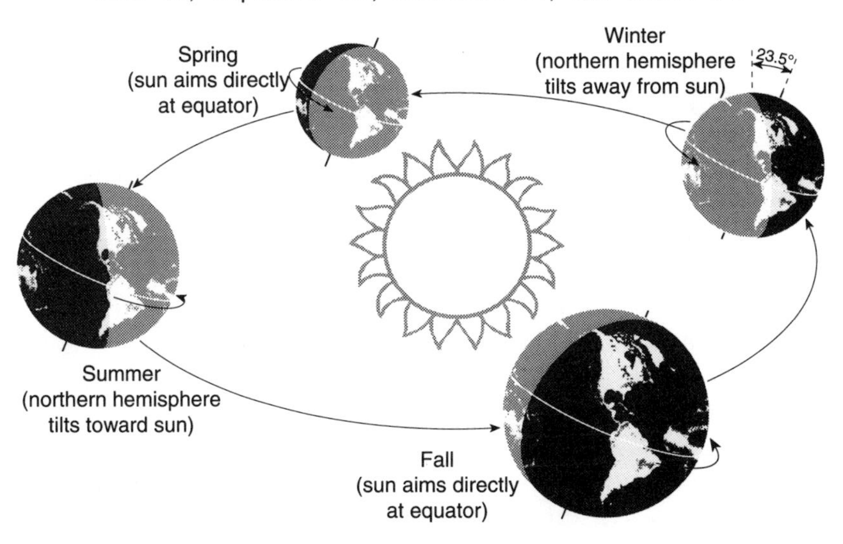

17. How is the weather different below the equator?

The earth has a 23.5° tilt with respect to its orbit around the sun (figure 5). This tilt results in the seasons, which alternate with each other above and below the equator.

During the northern hemisphere's summer the sun appears to be positioned north of the equator and daylight is longer. A date close to June 21 is the first day of summer for the northern hemisphere; in the southern hemisphere this is the first day of winter. Summer vacation for students "down under" typically occurs during the warm December-to-February weather. Contrasting with the north (Q. 16), the southern hemisphere is slightly closer to the sun during its summer and farther away in winter. This would cause southern seasons to be more severe than the northern hemisphere, but another factor arises: More water is below the equator, which moderates the seasonal change and gives a warmer southern climate year round.

18. What is seasonal affective disorder?

About 10 percent of adults are thought to experience this problem during the winter season. As the length of daylight shortens, they show symptoms of sadness, depression, and fatigue. In extreme cases the depression interferes with careers and relationships. The disorder is sometimes treated by exposing the patient to bright lights that artificially lengthen the daylight. People seem to have built-in biological clocks that are sensitive to light.

Seasonal affective disorder may be a new disease of our day. In earlier times, people more easily adjusted their minds and schedules to the changing length of daylight. In agriculture, the fall season brought harvest time and a change of responsibilities. Today, however, the typical full-time job continues through long summer sunlight and short winter days. Too often, the work schedule also includes the weekend. For some people, the result is seasonal stress and depression. One help is to build some variety into the seasonal schedule, based on the principle of a sabbath day of rest and worship. Planning or going on a vacation or a getaway weekend is good for people who are overstressed for whatever reason. And concentrating on the Lord's unfailing promise of the new season to come gives hope to one who feels hopeless.

19. What are the wind chill factor and comfort index?

The wind chill factor tells us how much colder the air *feels* when the wind is blowing. Cold temperatures are not as noticeable when the air is calm. A thin layer of warmed air next to exposed skin provides a measure of insulation. In a breeze, however, this air layer is constantly being removed and replaced by colder air. Table 1a gives some values of wind chill temperature. With a 20°F temperature, a 15 mile per hour wind gives a wind chill of –5°F. That is, a person loses as much heat as he would at –5°F in still air. Frostbite danger increases rapidly with wind speed, since heat is quickly removed from the body. Hypothermia, a lowering of the internal body temperature, is also a hazard. Wind chill does not affect the reading of a thermometer, since that device does not maintain a constant temperature as skin does. Wind chill also has no bearing on automobile antifreeze.

Wind chill factors combine temperature and wind speed, while the comfort index involves temperature and humidity. This is also known as humiture, heat index, or summer simmer index. In the summer, the comfort index measures how well our bodies can cool themselves. We lose most of our excess heat by direct radiation and evaporation of perspiration. Warm temperatures slow down the radiation heat loss; in addition, high humidity slows down evaporation from the skin. As a result we feel uncomfortable, as if the temperature were actually higher. Our bodies may feel tired, heavy, or "pressed in upon." How quickly we notice when our divinely designed cooling systems have difficulty. It is true that "it isn't the heat, it's the humidity" that causes discomfort. A dry heat, as experienced in most of the western states, is more bearable because the body can readily cool itself. Table 1b shows how humidity acts to increase the feel of the temperature.

You may have noticed that your hair is sensitive to moisture conditions. When humidity is high, the hair absorbs moisture from the air and becomes about 2 per-

Table 1a

Wind chill. Equivalent *colder* temperatures for various wind speeds, as felt by exposed skin.

Windspeed (miles/hour)	Air Temperature (°F)																
	35	30	25	20	15	10	5	0	–5	–10	–15	–20	–25	–30	–35	–40	–45
5	32	27	22	16	11	6	0	–5	–10	–15	–21	–26	–31	–36	–42	–47	–52
10	22	16	10	3	–3	–9	–15	–22	–27	–34	–40	–46	–52	–58	–64	–71	–77
15	16	9	2	–5	–11	–18	–25	–31	–38	–45	–51	–58	–65	–72	–78	–85	–92
20	12	4	–3	–10	–17	–24	–31	–39	–46	–53	–60	–67	–74	–81	–88	–95	–103
25	8	1	–7	–15	–22	–29	–36	–44	–51	–59	–65	–74	–81	–88	–96	–103	–110
30	6	–2	–10	–18	–25	–33	–41	–49	–46	–64	–71	–79	–86	–93	–101	–109	–116
35	4	–4	–12	–20	–27	–35	–43	–52	–58	–67	–74	–82	–89	–97	–105	–113	–120
40	3	–5	–13	–21	–29	–37	–45	–53	–60	–69	–76	–84	–92	–100	–107	–115	–123
45	2	–6	–14	–22	–30	–38	–46	–54	–62	–70	–78	–85	–93	–102	–109	–117	–125

Table 1b

Comfort Index. Equivalent *warmer* temperatures for increasing relative humidity. Hot temperatures in the lower right corner of the table mean maximum discomfort and danger of heat stroke.

Temperature (°F)	Relative Humidity (%)									
	10	20	30	40	50	60	70	80	90	100
70	71	71	72	73	75	76	77	79	80	81
75	75	76	78	80	82	84	85	87	89	91
80	80	82	84	87	89	91	94	96	99	101
85	85	87	90	93	96	99	102	105	108	111
90	90	93	96	100	103	107	110	114	117	121
95	95	99	103	107	111	115	119	123	127	131
100	100	104	109	113	118	122	127	132	136	141
105	105	110	115	120	125	130	135	140	145	151

cent longer. Dampness makes naturally curly hair more curly, and straight hair becomes even straighter. Some humidity instruments, called hygrometers, actually use the stretching of a strand of hair or hemp as a basis of measurement.

20. What is frostbite?

The freezing of one's nose, ears, fingers, or toes can happen without warning. The frozen parts become numb and pale colored. Since the bloodstream is actually 90 percent water, ice crystals may form during cold conditions. One serious result is the loss of circulation. Another is the expansion of water during freezing, which can destroy cells and damage skin tissue.

The recommended treatment for frostbite is gentle warming. The affected area should be covered with a blanket or another part of the body. Blisters may form later, similar to those of a burn.

Why are animals not frostbitten in winter? They have been provided with several different safety mechanisms for cold weather. Some have an insulated covering of fur, fat, or feathers. Small creatures usually have a rapid heartbeat, which helps them stay warm. The following list compares the pulse rates of some large and small cold weather animals with people:

Creature	*Pulse (beats/minute)*
Walrus	40
Person (resting)	70
Rabbit	200
Duck	240
Cardinal	400
Mouse	400
Shrew	800

Animals also have other means of protection from the cold. Birds, for example, have no blood flow in their feet. Some turtles, frogs, and fish develop an antifreeze additive for their blood. Researchers at Virginia Polytechnic Institute are studying synthetic versions of animal antifreeze. They seek new ways to prevent crops from freezing, noncorrosive substitutes for road salt, and coatings to prevent ice formation on aircraft wings.

21. Why is the sky blue, sunsets red, and clouds white?

The sky is indeed colorful, thanks to the sun. Its brilliant white light has all the rainbow colors embedded within: red, orange, yellow, green, blue, indigo, and violet (ROY G. BIV). When conditions are right, either with a rainbow or a sunset, the component colors become visible. The sky is blue because of the "scattering" of sunlight. The color blue has a shorter wavelength and greater energy than the other colors. As a result blue is selectively absorbed by air molecules, then given off again in all directions. The other colors are less scattered, and therefore not usually seen. The noon sun itself has a yellow appearance, having had its blue color subtracted out. At sunrise or sunset, the sunlight comes in at a low angle and must pass through a greater thickness of atmosphere. As a result the blue color is thoroughly scattered so that much of it is totally absorbed by air and lost. This leaves the other colors to be scattered, especially orange and red, making a glorious horizon of colors. These colors are further enhanced if there are dust particles in the air (Q. 5). The Creator's artistry far surpasses manmade displays.

The colors of sunlight are also responsible for all the hues we enjoy on earth, whether green grass or a goldfinch. The surface of each object selects the particular colors that it will reflect to distinguish itself. At night the creation assumes various shades of gray.

Clouds are often a brilliant white because they are excellent reflectors or scatterers of every color. All the returned colors together then add up to the neutral white. Certain common materials also reflect all the colors uniformly, such as milk, chalk, and sugar.

22. What is the ozone layer?

This is a thin layer of colorless gas in the upper atmosphere, twelve to eighteen miles high. Ozone (O_3) is formed when ultraviolet radiation (UV) from the sun interacts with normal oxygen (O_2):

$$O_2 + UV \longrightarrow O + O$$
$$O_2 + O \longrightarrow O_3$$
$$O_3 + UV \longrightarrow O_2 + O$$

The series of equations shows that ozone molecules are continuously formed and dismantled, with ultraviolet light being absorbed in the process. Ozone provides us with valuable protection against the ultraviolet radiation, which would otherwise hit the earth and cause a number of serious skin diseases. It can also damage our immune system. The ozone layer does not completely stop ultraviolet rays; thus caution is needed when sunbathing to prevent severe sunburn and perhaps skin cancer.

Ozone is actually quite rare. If the gas were concentrated down at the earth's surface, it would make a layer less than one inch thick. However, any deterioration of the protective ozone is of serious concern, and a slight loss has indeed been measured in recent years. Contributing factors include exhaust gases from jet airliners and solid fuel rockets. Other harmful chemicals are chlorofluorocarbon gases, or CFCs, that have been used in the past as refrigerants, cleaning solvents, aerosol sprays, and in foam insulation. One CFC example is Freon 12 with the chemical formula CCl_2F_2. These gases deplete ozone by interfering with the reaction equations listed above. Though CFCs are no longer used in great quantities, they have a long lifetime, perhaps a century. They therefore will continue to be a problem for a long time.

Seasonal "holes" in the ozone layer have also been noticed in the polar regions. These temporary openings may have a natural cause rather than being man- made, but they are getting larger. The ozone layer is provided by the Creator for our safety. Efforts to understand and preserve this precious part of the upper atmosphere should be encouraged.

23. How does the air handle pollution?

Much is said about the millions of tons of pollutants that enter the atmosphere each year: particulates, carbon monoxide, sulfur and nitrogen oxides, and others. However, we hear less about what eventually happens to all these chemicals. Throughout history the atmosphere has shown a surprising ability to restore and purify itself. This has been very important in the past, because natural events like volcanoes and forest fires often add much more dust and gas to the atmosphere than man does. Without this resilient ability of air to remain healthy, our air would long ago have turned into a brown mixture of poisonous chemicals.

Some pollution chemicals have well-known destinies: sulfur dioxide and nitrogen oxides form acid rain; much of the carbon dioxide is consumed by plants and trees. The weather also works to minimize problems by dilution of the air pollutants. Wind sweeps the contaminants away from their concentrated sources. Spread out, their consequences are less severe. It is when dilution and dispersion do not occur that acid rain and smog become serious problems.

In time, particulate material settles back to the ground by gravity. There it may become part of the soil. Many air pollutants are also removed by rain and snow, sometimes called washout. Solid and liquid particles stick to the moisture droplets. Meanwhile, some gases are dissolved in rain and cloud droplets; still other airborne chemicals break down into harmless elements through chemical and photochemical processes. For example, the reactive ion made of hydrogen and oxygen (OH) works to break down methane gas (CH_4). These natural cleansing processes, designed by the Creator, continually restore the atmosphere.

24. What makes a desert?

Desert has no exact definition. Generally it is a region with little precipitation, perhaps less than ten inches per year. Deserts range from the Arctic with its slowly accumulating snow, to northern Chile where it has not rained for four centuries. The deserts of our western states receive air currents that have already been squeezed dry by mountain heights. Such arid and semiarid regions cover nearly one-third of the earth's land surface.

The expansion of deserts into new areas is called desertification. It is taking place around the edges of Africa's Sahara Desert, in the Middle East, Australia, China, and the western United States. In poorer areas desertification results in economic hardship, malnutrition, and mass migration. Each year about 25,000 square miles of new land become deserts worldwide, an area one-half the size of Illinois. The advance of a desert occurs in an irregular fashion. It may halt during a wet period, then later move ahead quickly.

Desertification occurs when natural vegetation is lost and the exposed soil is then eroded by wind and water. This loss can be triggered in two ways. First, some areas simply experience a gradual trend toward a drier climate. (Continued climate adjustments in the post-Flood world are discussed in Q. 79.) Global warming would also have a drying effect on certain land areas. Second, human activities can accelerate desertification. Four particular activities have caused deterioration of land quality: overcultivation, overgrazing, firewood gathering, and overirrigation.

These activities often occur in heavily populated areas, especially in places that are already drought prone. The results are large tracts of barren land with hard, baked soil. Irrigation may eventually cause the soil to become salty and unproductive.

The biblical solution to desertification requires a worldwide awareness that we are responsible for nurturing the earth as stewards of God's creation (Q. 95). Putting severe ecological pressure on less-productive, marginal land only hastens its destruction. Instead, the use of land should not exceed its capacity to be stable. Areas adjacent to deserts provide buffer zones that need special protection. Whether the present world will realize this truth is uncertain. Someday, however, the desertification process will be supernaturally reversed. Deserts will once again become green and productive (Isa. 35:6).

25. Why do desert people often wear dark colors?

Bedouins are Arab people who live in the deserts of Jordan, Syria, and the Sinai Peninsula. They do indeed often wear black robes and they also have herds of dark-haired goats and sheep. This counters our usual view that white is a much better reflector of heat than black and thus would be a cooler covering to use in a desert. However, the Bedouin people are not in error. They have successfully lived in the desert for many centuries. Studies have shown that the heat gained by a person in a hot desert is actually the same, whether he wears a dark or light robe.

Additional heat is indeed usually absorbed by a dark surface color. However, greater convection or movement of air occurs beneath a black robe, which then carries the heat away before it reaches the skin. In fact, this chimney-like convection of air upward from the open bottom of the black robe makes the person feel more comfortable than a white robe. The same movement of air occurs beneath the hair of dark-colored animals; they remain healthy in hot deserts (see reference). In contrast, multiple layers of clothing in cold weather produce warmth by trapping air and preventing its motion.

The Sinai Desert can reach a temperature of 115°F at midday. Long ago, Moses led the children of Israel through this same wilderness when returning from Egypt. Like the present-day Bedouins, the Israelites probably learned the best desert survival methods during their forty-year tenure in this wilderness. It may well have included dark or colorful clothing instead of the white gowns often portrayed in illustrations of the Exodus.

26. Has the space program affected our weather?

Whenever the weather behaves strangely, it is tempting to look for a quick explanation. The space age has been especially vulnerable in this regard. After all, during the past four decades thousands of rockets have been launched upward through the earth's atmosphere. One naturally wonders about possible weather changes resulting from this continual disturbance. In truth, however, the launching of rockets has no measurable weather effects.

The atmosphere already contains the energy equivalent of trillions of rocket engines. Just one hurricane produces the power of a billion space shuttles. And while the space shuttle fires its engines for just a few minutes, a hurricane generates power for several days. The gases added to the air by rocket exhaust are likewise small when compared with other contaminants, both natural and man-made.

Actually the chief effect of the space program on the weather has been positive: more accurate forecasts. Weather satellites give us detailed pictures and data from around the world that we would otherwise miss.

27. How do weather satellites work?

Satellites have been helping us understand the weather since 1960. The pictures they send back to earth provide us with nature's own detailed weather maps. Some satellites have polar orbits, traveling north and south around the earth at a height of about 500 miles (figure 6). They circle the earth completely in about 100 minutes. In this way the entire earth gradually appears beneath them. Other satellites are geosynchronous or geostationary. They are located directly above the earth's equator at 22,300 miles high (figure 6). At this distance the satellites travel completely around the world in exactly 24 hours, identical to the earth's rotation time. These satellites thus appear to be suspended permanently above a given location on the earth, even though in reality they are orbiting at 7,000 miles per hour. One of these geosatellites can constantly "see" about one-quarter of the earth. A network of four such satellites, therefore, encompasses the entire world's weather.

Some weather satellites record infrared radiation rising upward from the earth. Their pictures clearly show temperature differences and the moisture content of the air. Other satellites take direct optical pictures. Either way, their pictures are sent to earth electronically, similar to a television signal. Weather satellites have been especially successful at detecting and following hurricanes over the vast oceans. No longer do these deadly storms take us by surprise. Satellite meteorology will continue to improve and show increasing detail, teaching us valuable lessons about the weather.

Figure 6

Polar (A) and geosynchronous (B) weather satellites. Most polar-orbiting satellites are about 500 miles from the earth; geosynchronous satellites are 22,300 miles from the earth.

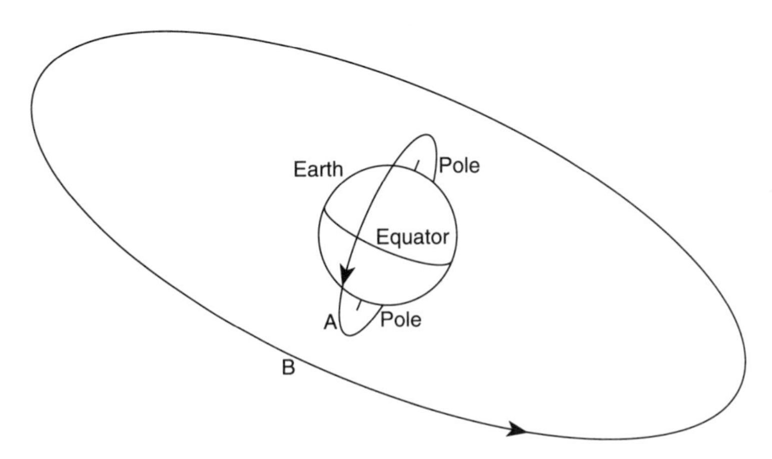

28. What is weather like on other planets?

It is a universal trait to complain about the weather. Conditions are seldom to everyone's liking. Perhaps a glimpse of the weather of faraway places would help us appreciate the earth. To that end, the following typical weather forecasts have been gathered from around the solar system:

The moon. No rain is forecast today; also no wind, clouds, or blue sky—it will be black instead. The high temperature this afternoon will be 215°F; the low temperature after dark will drop to –230°F.

Venus. Skies will remain overcast with heavy clouds today, just as every other day of the year. The surface temperature will remain at 900°F during the day and night hours. Thunderstorms are predicted, and air pressure remains high, 100 times greater than that of earth. Any humans venturing outside might be crushed.

Mars. Dust storms are approaching. Stay indoors since there is not enough outside air for humans to breathe. Lows tonight are expected to reach –100°F. All the planet's water has disappeared.

Jupiter. There is an extreme smog alert today. The air will be filled with poisonous hydrogen sulfide, also known as rotten egg gas. Be especially alert today for magnetic storms, falling meteors, and cosmic radiation. The "red spot" hurricane continues to roar, as it has for 300 years. This vast storm is large enough to swallow the entire earth.

Neptune. Fierce west winds will blow nonstop at 1,300 miles per hour today. Summer temperatures will hover around –325°F. It will be a dark day, since Neptune receives only one-tenth percent of the sunlight available on earth. Neptune's icy moon Triton is no better, perhaps the coldest spot in the solar system at –390°F.

Many of the solar system planets and moons have atmospheres, but not at all like our own. Instead, the measured gases are largely poisonous to humans:

Location	*Atmosphere*
Venus	Carbon dioxide (CO_2)
	Sulfuric acid (H_2SO_4)
Jupiter	Hydrogen (H_2)
	Methane (CH_4)
Titan	Nitrogen (N_2)
(a moon of Saturn)	Methane (CH_4)

The atmosphere of earth, even with its load of pollutants, is healthy and refreshing in comparison with our solar system neighbors.

29. What is solar wind?

There are many "winds" in space, but they are not like breezes on earth. They cannot cool the skin or make a flag ripple. Instead, space winds are fine sprays of small, atomic particles. They are a form of radiation and not much different from a pure vacuum.

The solar wind consists of hot particles ejected from the surface of the sun. During a "solar storm" the particles are especially abundant, mainly protons, electrons, and helium nuclei. As they speed through space much faster than bullets, some of these tiny particles reach the earth. Here the solar wind is deflected by the earth's magnetism toward the polar regions. As the particles collide with our atmosphere, the air molecules glow with energy. This gives the beautiful aurora, also called the northern (and southern) lights. Thus the solar wind is made visible in this beautiful display of color.

The aurora remind us of the Creator's care in protecting us from the solar wind radiation by his provision of a magnetic field around the earth. Without it we would be directly exposed to this harmful radiation. There is evidence that some stars give off a much more intense wind of radiation than the sun. These would be very dangerous regions to visit.

30. What is evaporation?

If a glass of water is uncovered for several days the water level usually drops. Liquid water is actually a swarm of rapidly moving and colliding molecules. Some of them gain sufficient speed to escape from the surface and travel into the air. Other water molecules already present in the air may likewise drop into the liquid. When more molecules escape than are captured, evaporation occurs. Above the surface of an evaporating glass of water, millions of tiny water molecules escape every second.

The large surface tension or "stickiness" of water ensures that evaporation is a slow process (Q. 33). Without this control, lakes would disappear and vegetation would quickly dry out and die.

It is the faster moving water molecules that are able to evaporate into the air. As a result molecules left behind have a lower average motion, and a slightly lower temperature (Q. 11). Evaporation is therefore a cooling process, sometimes noticed when moisture dries from your skin.

PART 2

Water, Wind, and Clouds

31. What is the water cycle?

The water cycle, or hydrologic cycle, describes the continuous movement of water throughout the environment. The journey begins with the vast reservoir of water in lakes and oceans. Heat from the sun causes some of this surface water to evaporate upward to form water vapor and clouds. The clouds eventually drop their moisture as either rain or snow, over both land and sea. On the land about 12 percent of the precipitation joins streams and rivers as surface runoff. The remaining 88 percent soaks into the earth as groundwater. This subterranean water moves slowly downward into the water table. Here it may be tapped by wells for human use. Groundwater may eventually appear as a spring, or it may feed into rivers, lakes, and oceans and possibly evaporate again. Solomon studied this water cycle three thousand years ago and described it accurately: "All streams flow into the sea, yet the sea is never full. To the place the streams come from, there they return again" (Eccles. 1:7).

Here are some of many interesting details in the hydrologic cycle:

As rainwater slowly soaks into or "percolates" through the ground, it is naturally purified for our use. Water flowing along a river is also cleansed unless the river itself is overwhelmed by too many contaminants. Water filtration and treatment plants in cities are artificial efforts to duplicate the natural cleansing process of groundwater and rivers.

Water that freezes in the polar regions of the earth may be removed from the active water cycle for a long period of time. Ninety percent of the world's freshwater supply is presently locked up in these polar ice caps. In the Antarctic, much of the ice reaches a depth of two miles.

On the land, water is transferred back to the air by both evaporation and transpiration. Transpiration is the loss of water from the leaves of plants and trees. Cells within the leaves must be moist to capture and dissolve carbon dioxide from the atmosphere. Through the process some of this moisture is then released into the air. This is also a necessary cooling process for vegetation, since evaporation cools. It is estimated that a single apple tree gives off 2,000 gallons of water to the air during one growing season.

One estimate gives approximately three thousand years as the time period for an "average" drop of water to move completely through the water cycle one time. Most of this time accumulates during the slow movement of groundwater, usually only a few feet per day. For this reason, pollution of groundwater is a very serious problem. Once in the ground, poison does not disperse quickly.

It was once thought that streams and rivers were replenished by underground channels delivering water directly back from the seas to the land. Around 1674, however, detailed studies were made of the precipitation falling in the collecting areas of rivers. It was found that there was enough rain and snow to account for all the water in a continually flowing river. Some of this precipitation flows directly into the river. Most of the water, however, sinks into the water table and feeds the river continually as it seeps out as springs. It gradually finds its way to the river even during long dry spells. This, then, is the permanent source of water for streams and rivers, all of which by gravity flow to the lowest places, the seas.

The total amount of water on the earth is estimated at 336 million cubic miles. That is 70 billion gallons for every person alive. It pours down on the earth at the rate of 1.5 trillion tons a day. A one-inch rainfall produces 65,000 tons of water per square mile, or 15 million gallons. The total distribution of water on earth is:

oceans	97.6%
icecaps, glaciers	1.9%
underground	0.48%
surface	0.02%
atmosphere	trace
total	100.0%

Even the "small" amount carried in the atmosphere as clouds and humidity amounts to 1,000 trillion gallons. The earth's water cycle is truly an immense, complex system.

The water cycle is obviously essential to our survival. It waters the earth, purifies the air, and cleanses the land. Meanwhile, no such cycle of water has been observed on any other planet. There may possibly be a form of precipitation on places like Titan, one of Saturn's moons. If so, however, the precipitation is probably droplets of liquid nitrogen, or perhaps snowflakes of methane, both at a frigid temperature near –200°F.

32. Why do lakes freeze from the top down?

Imagine a fictional world where ice sinks rather than floats. Broken water pipes or cracked radiators due to freezing would no longer be a problem, because ice would shrink instead of expand. Iceberg collisions with ships would also be unknown, since the ice formations would quickly sink out of sight to the bottom of the sea. However, new severe problems would soon become evident. With no floating layer of insulating ice, lakes would freeze solid from the bottom upward during winter. Thick ice layers would also form at the bottom of the polar seas. This subsurface ice would have devastating consequences on the earth's balanced hydrologic cycle (Q. 31). The solidly frozen lakes would thaw only partially during the summer season. Much less fresh water would be available for precipitation. The oceans would grow colder and more salty, probably killing most sea life. Human life would soon be endangered by this drastic ecological upset.

Fortunately, such a fiction is impossible. Everyone knows that ice floats, whether it is an ice cube or an iceberg. However, this special buoyant property sets water apart from all other common materials. Almost every other chemical compound becomes heavier when it freezes and then quickly sinks to the bottom of its liquid solution. The few chemicals that behave like water include the pure elements germanium, silicon, and bromine. The unusual lightness of ice is due to hydrogen bonding—the way that water molecules connect when they freeze (Q. 52). The resulting loosely spaced crystalline structure results in ice being seven percent less dense than liquid water, so the ice easily floats. The force caused by the expansion of freezing water is remarkable. Containers of freezing water will shatter, sometimes violently. New England farmers once split large granite boulders in their fields by boring small holes in the rocks and filling them with water. The confined water then froze at night and cracked open the rocks. The boulders could then be removed piece by piece. The expansion and floating of ice have many practical applications as well as provide a livable world. Ask an ice fisherman.

An intriguing reference to water is in Exodus 15:8. When the Israelites passed through the Red Sea, "the surging waters stood firm like a wall; the deep waters congealed in the heart of the sea."

A strong wind is said to have caused this separation of water (Exod. 14:21). The word congealed appears only this single time in Scripture. It describes a firmness given to the water, yet not a simple freezing. No such form of water is known in the laboratory. One may conclude that the Lord gives new properties to material at his command, entirely unfamiliar to us.

33. How does water moderate our weather?

Thousands of different chemical compounds are known, but common water is surely one of the most important to our well-being. Water is a very special resource; its physical properties are carefully matched with the needs of earth's land and inhabitants. Here are just a few properties of water that set it apart from other materials:

great stability	large heat capacity
high freezing point	high heat of vaporization
high boiling point	high heat of fusion
dissolving ability	high dielectric constant
large thermal expansion	high surface tension

To illustrate the unique behavior of water, let's briefly consider just one property, its large heat capacity.

Think about the baked potato you order in a restaurant. The outside aluminum foil is easily opened and removed with your fingers. But watch out for the potato itself. You must handle it with care or you will be burned. The food seems to hold more heat for a longer time than the wrapping does. In fact, the enclosed water, 90 percent of the potato by weight, has a "heat capacity" over four times greater than that of aluminum. This heat capacity is a measure of a material's ability to store heat energy. Water's heat capacity is one of the highest of any known liquid or solid in nature.

The large heat capacity of water has many applications. Practically, water is ideal for storing heat or transferring it from one place to another. Hot-water heaters, automobile radiators, power plants, and cooling towers all utilize this property. The large heat-reservoir property of water also is responsible for "evening out" the temperatures across the earth. Solar heat energy is readily absorbed by the seas and other surface waters, then given back to the air when air temperatures drop. Without this automatic thermostat and the atmosphere to distribute the heat, the earth's temperature fluctuations would be extreme, similar to the moon's. During the lunar day, surface temperature reaches 215°F. When darkness comes, the moon's temperature plunges to –230°F. Question 93 also explains that moisture in the air is a major absorber of the sun's heat, thus warming the earth.

The moderating effect of ocean water is readily seen by comparing Seattle, Washington, with Duluth, Minnesota. Both cities are at the same latitude, 47° north, but Duluth averages 37°F colder in the winter. Seattle is located just 100 miles from the warm Pacific Ocean, and thus experiences air warmed by the sea. Duluth has the northern plains to its west and experiences their cold winter winds. Consider also the Gulf Stream, which circles the North

Atlantic Ocean basin. It picks up an immense quantity of heat from the equator region and delivers it northward to Western Europe and the entire Arctic region. This vast current, equal to 1,000 Mississippi Rivers in volume, warms northern Norway to as much as 75°F higher than the average for this high latitude. In the Arctic, 150,000-ton icebergs can be melted in a single week by warmth from the Gulf Stream. The heat capacity of earth's water definitely controls the climate of the world.

Within our bodies the rapid movement of heat energy is essential. Water's large heat capacity helps the blood deliver warmth to the body's extremities. In this way our bodies maintain a constant, healthy temperature without any damaging cold or hot spots.

34. What causes the wind?

Wind is defined as air in motion. It moves from higher to lower pressure regions, much like water flowing downhill. During times of constant air pressure, the wind diminishes. The pressure differences themselves result from uneven heating of the earth by the sun. Most solar radiation is absorbed near the earth's waistline, in the region of the equator. The warmed air then moves both north and south. Along the way the air is deflected in different directions and to different altitudes. Wind thus works to even out differences in temperature and air pressure. The movement of large masses of air helps to distribute moisture across the face of the earth. Like a giant spoon, wind constantly stirs the atmosphere. Without it, severe hot and cold spots would persist at different locations.

The actual direction of wind depends on many factors including pressure differences, the spin of the earth (Q. 37), and surface features of the earth. The complex circulation patterns of wind are not yet fully understood (a common phrase in science). Winds are usually described as the directions they blow from. Thus, a west wind blows from west to east.

35. What makes the sound of wind?

Sound always originates as a vibrating object, whether a violin string, vocal chords, or a hanging leaf. The sound is carried to you by oscillating air molecules, which in turn set your eardrum vibrating. There are several ways in which wind makes itself heard. Trees provide a natural outdoor instrument. As branches and leaves shake, they cause vibrations in the surrounding air. The faster the object moves, the higher will be the pitch that is heard. Tree sounds are unique to those who stop and listen: the willow has been likened to a flute, the pine to a violin. High pitch sound also arises when narrow objects cause changing eddy currents of air, similar to a whistle. A lower pitch results when large open objects resonate, like air blown across a jug. These many effects sometimes make whistling, whirring, or moaning sounds, the combination of many frequencies. Air rushing directly past your ear itself can also resonate and play its music. We can tend to take these delightful sounds for granted or miss them altogether. However, they are an enjoyable part of the creation.

36. How is wind like the Holy Spirit?

This question brings to mind John 3:8:

> The wind blows wherever it pleases. You hear its sound, but you cannot tell where it comes from or where it is going. So it is with everyone born of the Spirit.

The word *spirit* in both Hebrew and Greek means "breath" or "wind." Both a breath of air and a breeze are appropriate images for the Holy Spirit.

Consider several properties of the wind. First, wind is moving air, and this fresh air is needed continually for life itself. Even seeds often require wind for their dispersal and subsequent growth. Similarly, the Holy Spirit is the presence of God, the source of all life. Second, wind has no material shape or form. It is invisible; we cannot see the source or the destination of wind. It is a mysterious, unseen force. Nevertheless, its presence is known by its effects. Likewise, the unseen Holy Spirit can be experienced in a refreshing way. His presence is displayed in the work he does in human lives by transforming, sanctifying, encouraging, and teaching. Third, wind is a powerful force. It cannot be stopped or controlled by people. Likewise, the Holy Spirit is not subject to human control. The moving of the Holy Spirit is God at work. Fourth, there is great variety in the wind. It may be a soft whisper gently rustling the leaves on the trees, or it may be a hurricane uprooting trees. Likewise the Holy Spirit may gently bring a person to Christ, such as a little child raised in a Christian home, or he may work in some climactic, dramatic way to bring conviction and conversion to the hardened sinner. In Acts 16, contrast Lydia, whose heart the Lord opened (v. 14), and the jailer, who needed an earthquake to jar him to his spiritual sense (v. 30). In both cases, the Holy Spirit did the regenerating work.

37. Why do weather systems have circular motion?

Satellite weather pictures show the continuous turning and churning of large cloud masses. This circular movement of weather patterns is actually caused by the spinning motion of the earth. At the equator the earth turns rapidly, about one thousand miles per hour. At the north pole, however, the earth has no spin speed at all. These locations are equivalent to the outside edge and center of a merry-go-round. Suppose you throw a ball while on a merry-go-round. To an observer, the ball will appear to follow a curved path. On the earth, moving air shows a similar deflection. At the latitude of the United States, air that would otherwise move north is turned to the east. This results in our familiar surface winds flowing from west to east—the "prevailing westerlies." In the tropics and polar regions this direction is reversed. There are similar air movements below the equator. This complicated curving of weather patterns is sometimes called the Coriolis effect, named for a French civil engineer. It also affects ocean currents, rockets, and airplanes. Job 37:12 may refer to the Coriolis effect on clouds: "At [God's] direction [the clouds] swirl around over the face of the whole earth to do whatever he commands them."

North of the equator, high and low pressure systems can be identified by their direction of turning. High pressure systems rotate clockwise and low pressure systems rotate counterclockwise. Severe low pressure storms such as hurricanes and tornadoes also spin counterclockwise over the earth's surface. South of the equator, these motions are reversed. Similar circular patterns have also been seen in the atmospheres of Jupiter and Saturn.

If the earth did not rotate, weather patterns would move in straight lines northward and southward from the equator. The warm equatorial air would rise, descending again near the poles. From there the cold air currents would return directly to the equator, giving the northern hemisphere a constant wind from the north. The world's weather patterns would be a drastic change from what they are, perhaps with particular "wind lanes" on the earth's surface where weather conditions would remain constant. Reduced mixing of the atmosphere would greatly increase environmental problems. The planet Venus, with its slow rotation of 243 days, does in fact have a direct equator-to-pole circulation.

38. What is the jet stream?

The jet stream is often shown on weather maps, bending to the north or south as it speeds across the country. It is roughly a ribbon-shaped river of air, several miles high, hundreds of miles wide, and moving from west to east at speeds of 50 to 300 miles per hour.

These high altitude winds were first discovered by bomber pilots during the 1940s. In flying westward toward targets in Japan, American planes accidently entered the jet stream and made little forward progress as they tried to "swim upstream" against the wind. Islands in the sea below them appeared to be stationary. It was soon realized that there are several of these "highways" of moving air that completely circle the earth. Pilots today avoid these channels of wind when flying west and often go with the wind when flying east. Flights from California to New York typically take an hour less because of this helping wind current.

The jet stream winds occur in the middle latitudes, some moving across the northern United States. They are associated with the "westerlies," a worldwide wind system that operates to partially smooth out temperature differences on the earth. The jet stream roughly divides cold polar air from the warmer air masses. The intense jet streams also act as steering currents for lower level weather systems. In the United States, the jet stream sometimes moves polar air southward and can also block the forward progress of storms. The effects on our weather are strongest in the winter. Over the months the jet stream appears to meander back and forth at random, like a waving ribbon.

Actually, of course, the movement of every weather detail follows the natural laws that the Creator has established for them.

39. What causes the mysterious circles in British fields?

Occasional reports describe the appearance of strange circular swirls of grain in farm fields. Corn and wheat are knocked flat in perfectly round patterns 30–300 feet across. Many explanations have been offered including pranksters, fungi growth, extraterrestrials, and midnight meetings of witches. In England, especially, hoaxers are proficient at trampling down grain by night. However, this does not completely solve the mystery. Records show that the unusual circles have been occurring for centuries. In the Middle Ages they were blamed on "mowing demons" who harvested fields after dark in dizzy circles.

The circular patterns have also been found in Europe, Australia, and America. They appear to be a weather phenomenon that requires a particular ground topography. One important factor seems to be an undulating golf course-like landscape, typical of Southern Britain. During a temperature inversion (Q. 12), a layer of cold air is trapped on the ground beneath warmer air. Under certain conditions, a column of warm air begins to spin downward. You may have seen similar small "dust devils" swirling across lawns or fields. This type of circulation, called an eddy vortex, may collapse and hit the ground. The air currents then sweep out a ring of slight damage as they expand, sometimes swirling corn stalks into a spiral pattern.

Eyewitnesses have described humming noises and light discharges as the circles form. Evidently dust or pollen particles in the spinning air can build up an electrical charge. This charge concentration may also burn areas of the circle, adding to the "flying saucer" interpretation of some circles. Genuine circles may be an example of the complexity of the earth's weather system. Weather details are complicated and unpredictable; however, extraterrestrial and demonic explanations are not needed. In the cases where circles show especially complex designs, their formation by pranksters is likely.

40. What causes clouds?

Clouds are visible moisture in the form of liquid droplets or ice crystals. The formation process may begin when warm, moist air moves upward from the earth's surface. The rising air expands outward as it encounters regions of lower air pressure. This cools the air in a process called adiabatic expansion. In a similar way, a refrigerator cools by compressing and then expanding a confined gas. Cool air cannot hold as much moisture as warm air. Therefore, tiny cloud droplets begin to form as the air rises. This process of condensation converts water from vapor to liquid. The droplets are "squeezed" from the expanding air, thus constituting the visible cloud.

Each round droplet of water produces a small amount of heat as it forms. All materials give up this latent heat when they change from a gas to a liquid. The resulting heat warms the cloud somewhat and limits further droplet formation. If the cloud rises higher, additional expansion, cooling, and droplet formation may then occur. By observing a cloud carefully, you can sometimes watch it grow on the edges as new droplets form and expand the cloud outward. This type of cloud is the familiar cotton-like cumulus variety. The cloud remains suspended in the sky because it is less dense than the air beneath it.

41. How are clouds named?

The different categories of clouds were first named in 1803 by an English weather observer, Luke Howard. The system has become quite complicated today with many subcategories of clouds. A description of a few of the general groups follows.

Cumulus. The name means a pile or heap. These are the fair weather, cottonlike clouds. They form from local, rising air currents. A typical small cumulus cloud may contain as little as twenty-five gallons of water, enough to fill a bathtub. If they enlarge greatly and tower high enough, giant thunderclouds may result. The water in these cloud systems may contain many millions of gallons of water.

Stratus. These are layered clouds that are stretched out like blankets. The flatness indicates that they are relatively undisturbed by wind. Gray stratus clouds may sometimes yield drizzle or snow.

Cirrus. Such clouds have a curly or wispy look like a lock of hair. They form two to five miles high, above the two types described above. At this high, cold altitude they consist of ice crystals. Cirrus clouds, sometimes called mares' tails, can be seen over a hundred miles distance. They often occur before an approaching moist, warm front with its accompanying precipitation. High cirrus clouds may also produce a ring around the sun or moon, sometimes predicting rain or snow.

Prefixes are often added to cloud names: *nimbus* for rain, *strato* for layers, *fracto* for fragment, *alto* for midlevel clouds. A thundercloud can thus be called cumulonimbus. These

Latin names of clouds are part of an international language used by meteorologists.

42. What causes fog?

Fog is simply a cloud that forms on the ground. It occurs when the air becomes saturated with moisture and droplets begin to condense. The droplets, a million times smaller than raindrops, are identical to those present in clouds. Cool air is heavier than warm air and it accumulates close to the ground in the evening. This cool air also holds less moisture than warm air. Evening fog occurs when water droplets condense from this surface air. Morning fog arises when there is additional cooling of the air around daybreak.

The fog first appears when the air is cooled to its "dew point." It also occurs above lakes and ponds, sometimes with the appearance of rising steam. At this time water is evaporating from the warm water surface and encountering cooler air above where it condenses. Along coastlines, fog often arises when warm moist air moves across cooler land or water. This advection fog often shrouds the Golden Gate Bridge in San Francisco. Cape Disappointment, Washington, averages about seven hours of such fog every day. Moist air exhaled from the lungs on a cold day cools quickly and also becomes visible as fog. In the winter, tiny suspended ice crystals sometimes occur, called ice fog.

Fog is a hazard to transportation; it can quickly bring airport traffic to a halt. This continuing problem shows the limited success of weather control. The fog problem is sometimes compounded in large urban areas when fog combines with smoke pollution to produce smog (Q. 12).

43. Why do jets sometimes leave contrails?

You have noticed the long white clouds left behind jet aircraft. Sometimes these contrails remain for several minutes, and at other times they don't appear at all. There are two reasons for contrail formation. First, the exhaust from burning jet fuel has water vapor as one of its components. When this vapor meets the cold upper air it immediately condenses into a trail of visible droplets. If the upper air is dry and stable, the droplets will quickly evaporate and disappear. If the air is humid, the contrail (*con*densation *trail*) evaporates less quickly. Second, if the jet exhaust particles freeze, they may provide condensation nuclei for moisture already present in the air. As a result, cloud droplets will form behind the aircraft. Their rate of dispersal also gives information on high altitude winds. A contrail that persists as a feathery cirrus cloud *may* indicate a moist warm front with approaching precipitation.

44. What are noctilucent clouds?

The name means "night-glowing" clouds. They are sometimes visible against the dark sky long after the sun has set. The clouds are located about fifty miles high in the sky, ten times higher than most ordinary clouds. At this extreme altitude the sun lights up the clouds long after sunset. They are too high and thin to be easily seen during daylight hours.

Noctilucent cloud particles, made up of dust coated with ice, are permanently trapped in the ionosphere. This is a low-temperature region of about –100°F, positioned above 99 percent of the atmosphere. Interestingly, these high-flying clouds were first reported in 1885, from Germany. Two years earlier, in 1883, a tremendous explosion had occurred on the volcanic island of Krakatoa, near Java,

Indonesia. Large amounts of water vapor and dust were thrown upward, much eventually reaching the ionosphere where it spread completely around the world. Volcanoes thus appear to be the major source of the noctilucent cloud material. The clouds also glowed brightly for several months after a large space object, either a meteorite or a small comet, hit Tunguska, Siberia, in 1908. This event also thrust a large amount of dust into the upper atmosphere.

PART 3

Stormy Weather

45. Is there a lull before a storm?

Often there is a brief lull in the air before a storm. It is caused by a balance between the forward movement of a storm and the winds being drawn into it. The lull may occur just ahead of a storm or directly below a thunderstorm before precipitation begins. With little breeze occurring the air seems extra quiet. Birds and insects may also become quiet as they adjust to the lower air pressure of the approaching storm.

In the case of a hurricane, the lull is within the storm. Since the hurricane is a large swirling mass of air, the relatively quiet middle region is like the center of a donut. This "eye" may be 10–15 miles in diameter; the barometer reading drops to its lowest point in the hurricane center. During the slow progress of such a storm, the eye may take 30 minutes to pass over a city. Car tires have been known to burst and corks pop out of bottles during such times. The strong wind from the leading edge of the hurricane subsides and the sky clears. Within the eye of the storm, towering clouds can be seen on all sides. This circle of clouds acts like a large chimney, sucking moisture upward from below. In the eye itself, the air sinks downward and warms, producing the cloud-free region. Flocks of birds are sometimes trapped within the storm's eye. As it passes, the wind picks up again in the opposite direction from earlier. When the hurricane moves on, the second half of the storm has begun.

46. Does Psalm 29 describe a storm?

Psalm 29 is generally recognized as one of the most graphic descriptions of a storm in ancient literature. David, the psalmist, calls the storm the "voice of the LORD" as it moves in from the north Mediterranean Sea and sweeps over the land. First we hear the distant thunder across the water, then we see nearby lightning flashes (vv. 3, 7). The storm shatters the mighty cedar trees of Lebanon and twists the great oaks. The mountains of Lebanon tremble before the strength of the storm (v. 6). The storm then moves on across the desert regions of the south with no letup in its intensity (v. 8). Verse 10 includes a flash flood resulting from the thunderstorm. The Flood of Noah's day may also be implied in this verse.

A storm is terrifying, but at the same time it is a majestic demonstration of power. The Lord is directly named eighteen times in Psalm 29; he is *in* the storm. God's people who visualize the universe as God's temple shout "Glory!" when they observe such a storm and give praise to God (v. 9). Try this praise the next time you experience a thunderstorm.

The Canaanite god Baal was closely related to weather and storms. Psalm 29 directly challenges the idolatry of Baal worship. It is God who is in control. Elijah also challenged Baal in calling down fire from heaven on Mount Carmel (1 Kings 18:16–46). Much later, on the Sea of Galilee, Jesus showed his deity by fulfilling a Psalm 29-type role: He stopped the storm (Mark 4:35–41).

47. What causes rain and snow?

Rain does not fall from "holes" in the sky, nor does it form by the "sweating" of clouds. Droplets that make up clouds are very tiny, ten times smaller than the diameter of a human hair. In contrast, the typical raindrop is roughly a million times larger in volume than a cloud particle. Therefore, a million cloud droplets must somehow coalesce or join together to form a falling raindrop.

In one process, larger droplets begin to form when water molecules stick to tiny "condensation nuclei" such as volcanic dust, smoke, pollen, or salt grains within a cloud. These growing droplets then fall through the cloud and sweep up additional moisture droplets, growing larger in the process. Finally they are heavy enough to leave the cloud and descend to the ground without evaporating completely. Updrafts in the cloud may help the process by increasing the number of particle collisions. Raindrops are limited in size by their tendency to disintegrate when they exceed about one-quarter inch. The surface tension of water is just the correct amount to prevent the drops from growing too large as they plummet to the ground. Otherwise they might be as large as bowling balls or pianos: Look out below! Raindrop size is also just right to disperse light into the rainbow colors (Q. 63).

A second process of precipitation formation occurs at higher levels in the atmosphere. Snow or ice crystals form at the subfreezing temperatures. The ice crystals then grow larger, consuming any nearby moisture in the process. Finally, the ice crystals clump together into snowflakes and drift to the ground. If surface temperatures are warm enough, the snowflakes will melt into rain during their downward trip. Some gentle summer rains begin as swirling snowstorms in the higher, colder levels of clouds.

48. What are biblical reasons for rain?

In Job 37:13 we find three distinct purposes of rainstorms. First, storms may be directed by God to punish people. The worldwide Genesis Flood is certainly the chief example.

Second, God sends rain to water his earth. This obvious benefit of precipitation should never be overlooked. If there were no rain, the surface of the earth would be like the moon's: dusty and lifeless. Even the size of raindrops falling on the earth is important. If they were larger, plant damage and accelerated erosion would result. If raindrops were smaller, like fog droplets for example, the rain would remain permanently suspended in the air. Surface tension, the density of water, and the strength of gravity all work together to provide the earth with pleasant showers of appropriately sized raindrops.

Third, according to Job 37:13, rainstorms show God's love. This is obvious from the first two points, since the discipline of people and the watering of the ground both demonstrate God's loving concern. In several other ways God's love is also shown by rain. A storm may provide the conditions for a rainbow, reminding us of the beauty of light and the important promise to Noah. A storm may also clean the air, providing refreshment and encouragement. God surely enjoys sending rain, thereby blessing his creation and helping it to prosper.

There may be still more reasons for rain that remain entirely beyond our understanding, and Christians should be cautious about searching for obvious, practical benefits in every detail of creation. The Lord may have his own hidden reasons. As Eliphaz reminded Job long ago, "[God] performs wonders that cannot be fathomed, miracles that cannot be counted" (Job 5:9).

49. Has it ever rained frogs or fish?

It may never rain cats and dogs, but fish, shells, and frogs have indeed fallen from the sky on rare occasions. In 1984, live six-inch flounders fell on a London neighborhood. It is thought that a waterspout had lifted them to cloud level from the Thames River, then later dropped them several miles away. The fish were probably as surprised as the people who saw them fall.

Similar stories are told of remarkable "showers" from other locations near open water. In one memorable storm in England in 1844, people held out hats to catch dozens of falling frogs. Such small creatures, once airborne, might be carried aloft for an hour or more within the strong updrafts of a thunderstorm. These stories remind us of the plague of frogs that occurred long ago in Egypt (Exod. 8). At that time, however, the frogs came onto the land from the rivers and ponds. This plague was much more extensive than a rare fall of frogs during a local storm (see reference).

Manna was an unusual, beneficial substance that God sent from the sky. This "bread of angels" (Ps. 78:25) gave nourishment to the Israelites during their forty years of wanderings in the desert. Exodus 16:14 describes the manna as "thin flakes like frost on the ground." Psalm 78:24 says God "rained down manna for the people to eat." Natural explanations of this manna have ranged from a honeydew substance made by insects to resin from trees. Numbers 11:7 says the manna indeed looked like resin, but it was clearly a miraculous gift from God (Deut. 8:3). It did not fall on the Sabbath, and stopped forever on the day that the Israelites began to eat the grain of Canaan (Josh. 5:12).

50. What is acid rain?

Either rain or snow can become acidic if it combines with chemicals in the air. Nitrogen oxides from automobile exhaust will convert water to a dilute form of nitric acid, HNO_3. Sulfur dioxide gas comes from the burning of coal and oil and turns moisture into a weak solution of sulfuric acid, H_2SO_4, the type of acid in a car battery. Acid rain usually is about 30 percent nitric and 70 percent sulfuric in its acidic composition.

The strength of acid is measured by a quantity called pH for its hydrogen concentration. Typical values of pH are:

oven cleaner	13.5
ammonia	11.4
sea water	8.0
pure water	7.0
acid rain	2.4–5.0
black coffee	5.0
orange juice	4.5
battery acid	0.5

The largest pH values represent bases or alkaline materials. A value of 7 is neutral. Notice that the smaller the pH value, the stronger the acid becomes. Precipitation has always been slightly acidic due to natural chemicals in the air. A single volcano may spew out thousands of tons of acid-forming gases. In recent years, however, the man-made portion of the problem has intensified. In Scotland, values of pH as low as 2.4 have been measured in rain. These are some of the many harmful effects of acid rain:

Acids wash nutrients from the soil and also damage root systems. Trees and plants turn yellow as a result. Weakened, the vegetation becomes vulnerable to disease.

Fish may die in acidic water, along with microorganisms and aquatic plants.

Brick surfaces and monuments crumble. The finish on autos becomes dull and pitted. The Statue of Liberty was restored in 1986 because chemicals had tarnished and etched its copper plating.

Respiratory problems increase when there are traces of acid in air moisture.

Acid precipitation illustrates the "limits of repair" that are built into the creation. Most soils, lakes, and streams have built-in basic chemicals that can react with acids to neutralize them. However, repeated doses of acid can overwhelm the defenses and break down the restoring process. Suddenly, large numbers of trees die or the water becomes sterile and devoid of fish. These occurrences remind us of our duty to care for the earth.

51. How does cloud seeding work?

Seeding is an attempt to increase precipitation amounts. A beginning requirement is the presence of clouds to work with. Ideally the top portions of the clouds should be supercooled. This means that liquid cloud droplets have not yet frozen, but the cloud's temperature is well below the freezing point. What is needed to trigger the formation of ice crystals is a large supply of freezing nuclei, or seed crystals. The chemical silver iodide (AgI) works well and is often dropped into the clouds by airplane. The tiny grains of AgI have a hexagon structure similar to ice (Q. 52). Thus the silver iodide triggers the formation of ice crystals which then rapidly grow larger. On the path to the ground, the ice melts into a rain shower. Since the resulting drops are a million times larger than the original AgI particles, the rain is not contaminated to any extent. Small pellets of dry ice, or frozen carbon dioxide, have also been dropped into clouds. These pellets cool the water droplets to below –40°F, where they spontaneously freeze and begin to grow.

Cloud seeding efforts thus far have had mixed success. Often the total rainfall is not increased at all. Instead the rain simply falls in the test area instead of somewhere else. This has led to lawsuits and charges of "rain stealing" by neighboring communities. On the other hand, the nation of Israel has been quite successful with rain modification. Experiments with both cloud seeding and reforestation in selected areas have shown a 15 percent increase in total rainfall. The Israeli government has an ongoing program to seed clouds at every opportunity. This has helped them turn their arid countryside green, especially around the Sea of Galilee.

52. What gives a snowflake its shape?

The outward form of a snowflake depends on its temperature of formation. The familiar six-pointed star shape occurs when cloud temperatures are between 3°F and 10°F. Above and below this range, tiny plates or pencil-shaped rods of ice will grow. The intricate shape of snowflakes can be partially understood from the chemical bonding mechanism of ice. As figure 7 shows, the sharing of hydrogen atoms between water molecules leads to a flat six-sided "chicken wire" pattern in ice crystals. A snowflake consists of hundreds of these tiny ice crystals, with trillions of water molecules included, and reflects the beautiful inner hexagonal symmetry.

The growth and variety of ice crystal shapes in nature was studied by Robert Boyle in 1611 and Robert Hooke around 1700. They concluded that crystals such as snow "receive their regular figures from the texture or nature of their own parts." Today, active study continues on the beautiful pattern of snowflakes. The exact shape depends on conditions occurring right next to the tiny flake: air temperature, wind current, humidity, nearby snowflakes, etc. We can understand why no two flakes are completely identical, either on the outside or inside, since each micro-environment is unique.

Figure 7

The hexagonal crystal structure of water in the form of ice. The H and O symbols are hydrogen and oxygen atoms. Additional hexagons can be constructed around the edges.

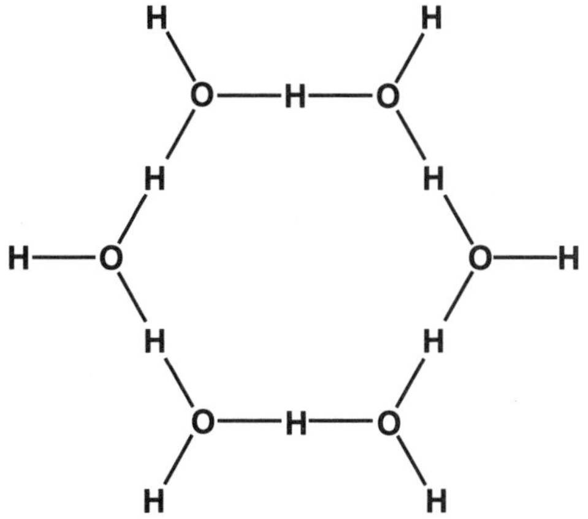

53. Is every snowflake different?

Incredible as it may seem, every snowflake that has ever fallen must be unique. Since the total number of snowflakes is so vast, statistics must be used in any analysis. In just one cubic foot of snow are about 20 million flakes. During all of history, between 10^{30} and 10^{40} snowflakes have probably formed. In contrast, the total number of known stars in the universe, 10^{22}, is a billion-

billion times smaller than 10^{40}. Now a basic question is, How many different snowflakes are possible? A single flake consists of perhaps 10^{18} water molecules. They are bonded together in the familiar, hexagonal snowflake arrangement (figure 7). Even when two snowflakes appear identical on the outside, multiple variations within the flakes nevertheless will make them quite different. Every snowflake contains a unique combination of dust particles, impurity atoms, and imperfections in the ordering of atoms. If just ten such variations are considered anywhere within a snowflake, then the typical flake with 10^{18} water molecules can be arranged in almost endless different ways. This is somewhat similar to having a stadium filled with thousands of people, and then calculating the vast number of possible seating arrangements. If any two or more people exchange seats in any way, it counts as a new arrangement.

The number of possible variations within a single snowflake can be estimated. Assume a collection of r water molecules and s impurities. There will then be

$$N_{r,s} = \frac{r!}{(r-s)!\,12}$$

unique possible arrangements. (The exclamation symbol means factorial.) For example,

$$\begin{aligned} 4! &= 4 \times 3 \times 2 \times 1 \\ &= 24 \end{aligned}$$

The 12 in the denominator arises from the sixfold rotational symmetry of a snowflake and from the fact that a flake can be looked at from either side. If $r = 10^{18}$ molecules and $s = 10$ distinct impurities, then the number of arrangements is

$$N_{r,s} \sim \frac{(10^{18})^{10}}{12} = \frac{10^{180}}{12}$$

Given 10^{40} total snowflakes over all of history and $(10^{180})/12$ possible variations, the probability of having two identical flakes is about 6×10^{-100}, an incredibly small number. Suffice it to say that the chance of exact snowflake duplication is zero (see reference).

By similar reasoning, every sand grain and blade of grass, every person and star is uniquely different. And we shouldn't be surprised. The creation demonstrates the infinite ability of God to design and also care for what he has made. God performs wonders that cannot be fathomed (Job 9:10).

54. What are some useful properties of snow?

Snow blankets large regions of land each winter season, changing the landscape and often closing roads and schools. Since both water and snow-forming temperatures are very rare in the universe, falling snow may be a unique blessing to the earth, unknown anywhere else. Among the many benefits of snow, four will be mentioned here.

First, snow provides a slow, measured supply of water to underground water reservoirs. As snow slowly melts, the liquid percolates downward into the water table. This is especially true of large snow accumulations at high elevations. In contrast, rainwater is often quickly carried away as surface runoff, with accompanying erosion. As everyone knows, a permanent supply of fresh water is essential to life, and snow is an especially suitable provider.

Second, a snow cover insulates the ground during cold weather. Studies have shown that nighttime temperatures can be more than twenty degrees warmer beneath a layer of protective snow. This insulation keeps the ground from freezing to great depths, an important safeguard for the water table that prevents widespread permafrost. A blanket of snow also protects small animals such as mice, which must remain active during the winter. They often construct a maze of tunnels under the snow. Such creatures are part of the food chain, and their winter survival is necessary to many other creatures.

Third, falling snow is an excellent purifier of the air. Every snowflake captures floating dust particles as it begins its growth. As the snow falls, it further sweeps up other contaminants in the atmosphere. Both snow and rain thus provide us with fresh air. The whiteness of snowflakes is a suitable metaphor for God's cleansing of our sin (Isa. 1:18).

Fourth, it is true that every snowflake is uniquely different (Q. 53). In this way snow shows the creative glory of God, just as it is also seen in every other aspect of nature. The heavens declare God's glory (Ps. 19:1), and so do snowflakes.

55. What causes thunder and lightning?

Lightning begins with the separation of positive and negative electrical charges in a cumulonimbus or thundercloud. Water droplets move upward and downward within the cloud because of updrafts and gravity. These droplets collide with each other and also with ice crystals. A separation of charge results, similar to the frictional effect of a person walking across a carpet and becoming charged with static electricity.

Lightning sometimes occurs in winter, but ice crystals and snowflakes are less likely to produce a charge separation. In rare cases a lightning bolt occurs "out of the blue" with no storm clouds present, when charge separation must have unexpectedly occurred in a normal overhead cloud. Benjamin Franklin first flew his kite during a Philadelphia thunderstorm in 1752. He noticed that sparks jumped from a key tied to the kite string, to his knuckles. This *dangerous* experiment showed that lightning is indeed electrical in nature.

When the negative electron charge in a cloud builds to a critical amount, an invisible avalanche of electrons rushes to the ground. With this electrical channel opened, additional electrons can then flow from the cloud to ground with little resistance. Electrons at the bottom of the channel drain out first, followed by successively higher positions. This "return stroke" causes the visible lightning, which actually moves from the ground upward. The return stroke carries perhaps 100,000 amperes of electricity at 30 million volts. In a fraction of a second, the lightning may streak back and forth several times. It may also remain in the air, jumping between clouds. Lightning is thus nothing more than a giant electrical spark.

The actual path of a lightning bolt appears to be crooked and haphazard. However, Job 38:35 reminds us of the Creator's detailed control of nature. He even opens up the conducting path for the electrons: "Do you send the lightning bolts on their way? Do they report to you, 'Here we are'" (Job 38:35)?

Thunder results from the intense heating of air by the lightning. A lightning stroke generates an instant temperature of about 50,000°C, eight times hotter than the surface of the sun. This heat energy is quickly transferred to the surrounding molecules of air. A shock wave of colliding molecules then moves outward from the lightning source at the speed of sound, about 750 miles per hour. This pulse of high pressure is actually a sound wave which eventually reaches us and causes our eardrums to vibrate. If all the sound arrived at once, we would hear a loud explosion. Instead, the length of a lightning stroke across the sky causes the thunder to rumble for several seconds as the sound arrives from the increasing distance.

The time delay between a lightning stroke and its resulting thunder can be used to estimate the storm's distance. The flash of light is seen almost instantly, but thunder (sound) travels more slowly than light, about one mile every five seconds, or one kilometer in three seconds. If the time delay between your seeing the lightning and hearing its thunder is ten seconds, for example, then the lightning occurred two miles away. You can measure the number of seconds by counting "one thousand, two thousand . . ." between the flash and the thunder. If the thunder arrives in just a split second, then the lightning was very close. If a lightning flash is more than ten miles away, its thunder is rarely heard. This type of lightning on the far horizon is popularly called heat lightning. However, it is actually part of a normal, distant storm.

A thunderstorm is beautifully described in Job 37, where thunder is likened to the voice of the Lord (Job 37:2). At any given moment, it is estimated, at least 2,000 thunderstorms are occurring somewhere around the world, but mostly in tropical regions. Only the infinite Creator could be in control of such detail. The sound and fury of a storm is naturally frightening to us. We are impressed by power, and any glimpse of the power and majesty of God can be unsettling. However, a lightning storm beautifully displays God's glory, and it also has practical benefits for the earth (Q. 56).

56. What is lightning good for?

In the creationist view of science, every detail of nature displays God's planning and purpose. The cosmos is not just the result of time and chance. Of course, we must avoid the attitude of demanding to know, as a condition for faith, what every single detail of nature is good for. Many parts of God's handiwork remain beyond our understanding. The creation declares God's glory, and that alone is sufficient reason for its existence. Still, science continues to discover the value and interrelationships of created things.

A lightning stroke provides a good example of purpose in nature. Lightning may frighten us, but it also serves an important function: It helps provide our daily need of the element nitrogen. Nitrogen is the third most abundant element in the human body and it must be renewed continually. We are actually surrounded with nitrogen, since it comprises 78 percent of the earth's atmosphere. However, we cannot use the nitrogen in this molecular, gaseous form; nitrogen gas has a strong covalent bond that our bodies cannot break down. Instead, nitrogen must first be taken up by plants as part of the food chain.

There are two major ways in which nitrogen becomes a part of vegetation. First, certain plants are able to absorb or "fix" nitrogen through their root systems. These plants are called legumes and include clover and peas. Legumes are distinguished by nitrogen-fixing bacteria that live inside their root nodules. These bacteria change the N_2 form of atmospheric nitrogen into nitrate, NO_3:

$$N_2 + 3O_2 \longrightarrow 2NO_3$$

In the nitrate form the legumes, as well as other plants, can utilize the needed nitrogen. Through a diet that includes either these plants directly or the products from grazing animals we make the nitrogen available to our bodies.

The second major way by which nitrogen is transformed into a usable form is by lightning. The great electrical energy of lightning is easily able to convert N_2 to NO_3. The magnitude of this lightning transformation process is remarkable. Worldwide, storms are estimated to produce one hundred million tons of useful nitrogen compounds each year. Recent studies indicate that up to one-half of all usable nitrogen in the soil is fixed or made usable by the energy of lightning. In our day of intensive agriculture, farmers supplement the large natural supply of nitrogen by adding ammonia (NH_3) to their fields. At the same time, the natural enrichment of soil by lightning and legumes continues, day by day. The next time lightning is crackling across the sky, remember to thank God for his provision of our essential nitrogen.

57. Can lightning strike the same place twice?

Yes, lightning can indeed be attracted to the same object repeatedly, either during a single storm or at a later time. Lightning is caused by an electrical charge buildup between a cloud and an elevated ground object. A tremendous attractive force can result between the positive and negative charges. The charges are originally separated within a cloud by the collision of water droplets and ice crystals. Immediately after a lightning stroke occurs to equalize the charge difference, charge separation may start to rebuild again. In this way, for example, the tower above New York's Empire State Building is often struck by lightning. Each time, the electrical charge is harmlessly carried into the ground.

The *Guinness Book of Records* describes a man who was struck by lightning four times over a thirty-year period.

Lightning is very energetic and unpredictable in its effects. Some people who stand near a lightning bolt are fatally injured; others can actually be struck and yet largely be unharmed. Any electrical current that moves within the body, especially near the heart, can be a great danger. However, lightning has such an intense current that it sometimes moves along the outside of a person's skin without entering the body. The result may be little more than a severe burn.

To minimize lightning danger during a storm, avoid standing out in the open or under a lone tree. Swimming and boating are also hazardous. There is relative safety inside a house or building. On rare occasions a shock may be received through the telephone; it is therefore advisable to delay nonessential calls until an intense storm has passed.

58. Do certain trees tend to attract lightning?

Actually, all trees attract lightning. They are tall and usually filled with moisture, thus providing a good electrical conducting path for lightning. Every day, thousands of trees are hit by lightning throughout the world. Past studies of the kinds of trees struck are complicated. It depends on tree height, the variety of trees present, and soil conditions. Overall, the oak tree appears to be most vulnerable: "Beware the oak; it draws the stroke." Oaks tend to be taller than the surrounding trees, thus attracting the lightning. They also have a high moisture content, which increases their ability to conduct the surge of electricity.

When a tree is hit by lightning, most of the electrical charge moves through the outer, growing portion of the tree. This intense current can instantly vaporize sap into steam. As a result, the tree may violently split or even explode. About half of all trees struck by lightning survive for the moment. However, they are often weakened and made susceptible to future disease. A target tree is better off if struck after it has been thoroughly soaked by rain. Much of the electrical charge is then able to move safely down the outside surface moisture instead of through the internal tree.

59. What is ball lightning?

Actually there are several unusual forms of lightning. Ribbon lightning occurs when a strong cross wind spreads out the flashes of a multiple-stroke flash. It then looks like a double or triple exposure. Sheet lightning is the brightening of an entire cloud by internal flashes. In bead lightning the stroke is broken into several distinct bright segments. Finally, ball lightning is the rare formation of a glowing sphere, often about one foot in diameter. The bright red or yellow ball may move along the ground with a lifetime of several seconds. Its makeup has been likened to a plasma, the glowing gas inside many street lights.

Ball lightning shows very peculiar behavior. On one occasion the ball entered a home through an open window, bounced on the carpet, and then disappeared into an electrical wall outlet. Another ball, glowing like a fluorescent light, disappeared into a tub of water, which then boiled for several minutes afterward. Ball lightning may even provide an explanation for some of the reports of unidentified flying objects (UFOs).

The existence of this unusual lightning reminds us that many details of the weather are not well understood. There are indeed a variety of unidentified objects, but many are weather related and none are Martians. Perhaps some of the atmospheric judgments on the earth, such as the fire which rained down on Sodom and Gomorrah

(Gen. 19:24), included divinely appointed concentrations of ball lightning.

60. What is Saint Elmo's fire?

This strange term is given to a visible glow that sometimes appears at night on objects that are charged with electricity. It may have a green color and make a crackling noise. More scientifically it is called a corona or electrical discharge. The original name is derived from Saint Erasmus, an Italian bishop in the fourth century. He was known as the protector of sailors on the Mediterranean Sea. During lightning storms a glow would sometimes appear above a ship's masts and was thought to show the patron saint's presence. The glowing fire above ships was also associated with Castor and Pollux, two Greek deities. Acts 28:11 records that the apostle Paul sailed to Rome in a ship with the figurehead of the twin gods Castor and Pollux. Paul experienced his share of violent storms at sea and likely was familiar with the strange glow of light on ships' masts. He probably told his shipmates that the Lord was in control of storms, not Greek gods.

Coronal discharge or Saint Elmo's fire is sometimes visible on the sharp edges of aircraft. As the plane moves through air it becomes charged by friction. The glow is static electricity and is not a particular hazard to air travel. When the plane lands, it is usually "grounded" with a wire cable to drain off any remaining charge. Some aircraft also have a grounding wire in the landing gear. During an evening thunderstorm, you may notice the faint glow of an electrical discharge around electrical wires or above a lightning rod. Mountain climbers have also reported green halos of light around their ice axes, and even around the horns of mountain cattle.

A Bible critic once suggested that a natural coronal discharge was responsible for the bush that burned before Moses without being consumed (Exod. 3). Of course, God is capable of carrying out his miracles in ways beyond our understanding. However, electrical activity does manifest itself throughout the creation.

61. What happens to thunder?

Thunder results from the vibration of air molecules. Sound is always a vibration, whether that of a guitar string, an air molecule, or an eardrum. This vibrating motion is a form of kinetic energy. As a sound wave spreads out from its source, the vibrational energy is shared among increasing numbers of air molecules. Soon, the sound energy is diluted and weakened below our threshold of hearing. However, the vibration energy is always "conserved" and does not disappear. Instead, the sound energy from every thunderclap contributes to the continuous "background" motion of all atoms in the air, in water, and even in the walls of a house.

The sound of a human voice provides an interesting extension to the initial question. The past vibrations caused by a particular voice, which is always unique, still remain present in the background of random atomic motion. Of course, it is technologically impossible to recapture a particular voice sound from the past, because the energy is gradually shared among atoms all over the world. Still, it is true that our very words from former years are still around in a real, physical sense, mingled with everyone else's. Lincoln's 1863 Gettysburg Address also is still "in the air" somewhere.

This sobering thought is also true of pictures from the past. As the sun lights the earth, an image reflects off people and objects. A camera may capture an instant of this image, but the entire moving picture scene travels continually outward into space at the speed of light. This light motion is different from sound, which cannot travel beyond the earth's atmosphere. One can imagine a super telescope, built with perfect seeing ability, pointed at the earth from a distance of 100 light years, or 588 trillion miles away. This instrument would see events on the earth happening just as they actually did a century ago. There is thus an actual, visual record of earth events speeding outward into space at all times.

We are reminded of the book of life, described in Revelation 20:11–15, which records the deeds done in this life. Some have questioned the idea of every past detail being recorded and remembered. However, now we can understand that there actually is a real, physical record of both our words and deeds from the past. Keeping this permanent, growing record of our lives in mind might result in more disciplined living. Of course, the Lord is above time, and can surely see and hear all things in a perfect way that is entirely beyond our comprehension.

62. How do lightning rods work?

The lightning rod was first described by Benjamin Franklin in 1753. He wrote about his invention in *Poor Richard's Almanac:* "It has pleased God in his goodness to Mankind, at length [to reveal] to them the means of securing their habitations and other buildings from mischief by thunder and lightning."

The technique is to attach pointed metal rods to the highest points of a building. The rods are then connected by wire cable to metal stakes driven into the ground. If electrons surge from a thundercloud to the house, they will take the path of least resistance from the lightning rod down through the cable and into the ground. Thus the house will be spared from damage and possible fire. Lightning rods do not actually prevent lightning but safely divert the resulting electrical current.

When a house without lightning rods is struck, the charge often follows water pipes or wiring into the ground. It may burn out circuits or shock someone in contact with water or a telephone. Lightning can also destroy a chimney or start a fire.

In 1769, forgetting past experience, leaders in Brescia, Italy, declared it sacrilegious to suggest that God would strike a church with lightning. They therefore refused to mount lightning rods on churches. Instead, gun powder was stored in churches as the safest possible arsenals. Soon thereafter, a church steeple was struck by lightning, and 3,000 people died in the resulting explosion which leveled the city. The value of lightning rods was quickly realized by everyone.

Buildings with steel frameworks are usually safe during lightning. If struck, the metal provides a safe electrical passage into the ground. Likewise, a car is quite safe, because its metallic outer body protects the occupants. Electricity tends to flow on the outer surface of metallic objects. Tires also help insulate a car, just as rubber-soled shoes provide relative safety for an electrician. The outer metal of ships is a natural lightning rod, and lightning is rarely a hazard. Nonmetallic boats can be outfitted with lightning rods, with the grounding wire connecting to an underwater metal plate. Airplanes are less vulnerable since they are not in contact with the ground (see reference).

63. What causes a rainbow?

A rainbow occurs when raindrops and sunshine cross paths. Sunlight consists of all the colors of light, which add together to make white illumination. When sunlight enters water drops, it reflects off their inside surfaces. While passing through the droplets, the light also separates into its component colors, which is similar to the effect of a glass prism. Each falling water drop actually flashes its colors to the observer for just an instant, before another drop takes its place.

A rainbow is usually seen in the opposite direction in the sky from the sun. The rainbow light is reflected to the eye at an angle of 42° to the original ray of sunlight. The bow shape is actually part of a cone of light that is cut off by the horizon. If you travel toward the end of a rainbow, it will move ahead of you, maintaining its shape. Thus there is no real end to a rainbow, and no pot of gold waiting there. Because the 42° angle is measured from each individual observer's eye, no two people see exactly the same rainbow. Every person is at the center of his or her own particular cone of colored light. From the high vantage point of a mountaintop or an airplane a complete circle of rainbow light sometimes can be seen.

The bright, primary rainbow has red on the outer edge and blue within. Higher in the sky there is always another, dimmer rainbow with the order of colors reversed. This secondary rainbow results from additional reflection of sunlight through the raindrops. It is most visible when there are dark clouds behind it. Look for the second bow high in the sky the next time rainbow colors appear. Some observers have even reported seeing third and fourth rainbows above the first two.

The technical details of rainbow formation were first analyzed by Isaac Newton in 1665. His brilliant optics work concerning reflection and refraction certainly does not detract from the beauty and promise of the rainbow. On the contrary, Newton's scientific insights show the marvelous complexity of the creation. The rainbow is a gracious pledge that God will not destroy the earth a second time with a worldwide flood (Gen 9:11–17):

> I have set my rainbow in the clouds, and it will be the sign of a covenant between me and the earth. . . . Never again will the waters become a flood to destroy all life (Gen. 9:13, 15).

64. What makes a squall line?

Perhaps you have watched an ominous straight row of dark thunderclouds approaching just before a storm. This squall line usually forms in front of a rapidly moving cold front. The advancing cool air forces warm, moist air ahead of it to rise. This rising air causes brief heavy rain and wind. The squall line will sometimes extend for hundreds of miles in length. It is both frightening and awesome to see the power that can quickly arise in our sky. Actually, a squall line is a blessing in disguise. The line of turbulence acts to defuse instability that might otherwise attain far more destructive power.

65. What causes chinook or Santa Ana winds?

These are very warm dry winds associated with mountainous areas. First, warmer air is forced up the windward side of a mountain range. At higher elevations it cools and loses its moisture by precipitation. As the dry air then descends on the leeward, downslope side of the mountain, it warms by compression and also picks up speed, to as much as 125 miles per hour. Along the eastern slopes of the Rocky Mountains the wind frequently goes by the name of chinook, an Indian name from that region. It is also called a snow eater wind, because snow lying along the path is rapidly melted or evaporated. The dehydrated wind also dries out vegetation and may fan raging brush and forest fires. The drying effect crumbles leaves, scorches corn, and can practically bake apples hanging on trees. In Rapid City, South Dakota, a chinook racing out of the Black Hills once increased temperatures from –2°F to +47°F in just five minutes. This extreme temperature change gives new meaning to the name Rapid City!

In the European Alps the name *foehn*, meaning wind, is given to the phenomenon; it goes by the name Santa Ana in southern California. The strong winds have been known to overturn trucks on California freeways and derail trains in Montana. Hot, dry winds are also described in Scripture. Jeremiah 4:11 describes a scorching wind coming down from the desert mountains. Ezekiel 19:12 also tells of oppressing east winds that dried fruit on the trees, just as they still do in Israel.

66. Why does the Sea of Galilee have sudden storms?

A storm on the Sea of Galilee is described in Mark 4:37: "A furious squall came up, and the waves broke over the boat, so that it was nearly swamped." Luke 8 further describes the wind and raging waves during this storm. Matthew 8 calls it a furious storm without warning.

Such storms result from differences in temperature between the seacoast and the mountains beyond. The Sea of Galilee lies 680 feet below sea level. It is bounded by hills, especially on the east side where they reach 2000 feet high. These heights are a source of cool, dry air. In contrast, directly around the sea the climate is semi-tropical with warm, moist air. The large difference in elevation between surrounding land and the sea causes substantial temperature and pressure variations. This results in strong winds descending to the sea, funneling through the hills. The Sea of Galilee is small, and these winds may descend directly to the center of the lake with violent results. When the contrasting air masses meet, a storm can arise quickly and without warning. Small boats caught out on the sea are in immediate danger.

The Sea of Galilee is relatively shallow, just 200 feet at its greatest depth. A shallow lake is "whipped up" by wind more rapidly than deep water, where energy is more readily absorbed. Lake Erie provides somewhat of a parallel with the Sea of Galilee. Erie is more than a hundred times larger, but it has the same 200 feet maximum depth, the shallowest of the Great Lakes. Lake Erie is especially well known as the stormy, moody member of the Great Lakes system. It is easily stirred up by west winds to produce violent waves and even the largest fishing boats are put at risk.

67. Are there storehouses for the hail?

This question is based on Job 38:22, 23: "Have you entered the storehouses of the snow or seen the storehouses of the hail, which I reserve for times of trouble, for days of war and battle?" The wind is also described in Scripture as issuing forth from God's storehouses or treasuries (Ps. 135:7; Jer. 10:13; 51:16).

As far as we know, there are no vast reservoirs of wind, prepared snow, or hailstones in the earth's atmosphere. However, the ingredients are surely present. Almost always abundant moisture is in the air. Clouds themselves consist of either liquid droplets or ice crystals. When conditions are right, by God's direction snow or hail will readily form.

Hail results when the storm clouds are especially turbulent. If they are at a high, cold altitude, ice crystals are present, even if the ground below is warm. The ice crystals are carried downward within a cloud by gravity, then back upward by strong wind updrafts. As the ice crystals move upward and downward, they become coated with multiple layers of ice. A broken hailstone looks somewhat like a small onion with many inside layers. The largest recorded hailstone in recent times fell in Bangladesh in 1986. It measured six inches across and weighed 2¼ pounds.

An even greater hailstorm helped Joshua defeat the Amorites: "The LORD hurled large hailstones down on them from the sky, and more of them died from the hailstones than were killed by the swords of the Israelites" (Josh. 10:11).

The truth of the Job passage regarding storehouses for hail is that energy, material, and a mechanism are all available for the formation of snow or hail. At God's command they will appear in abundance.

Hail is a symbol of God's judgment, along with fire or lightning. These two destructive ingredients fell together on Egypt during the seventh great plague (Exod. 9:23, 24). This passage declares that there had never been such a storm in Egypt's entire history. From Revelation 16:21 we learn that a cleansing storm of fire and hail is also forecast during the last days (Q. 99). These hailstones will weigh a talent, or about 100 pounds each.

68. Why is a downburst dangerous to aircraft?

In 1985, a commercial airliner was slammed to the ground at the Dallas airport, resulting in 133 fatalities. Suddenly, the downbursting of air became a sad reality.

A downburst or microburst appears to be a rapid pouring downward of cold air from a cloud. Like water from a faucet, the cold air hits the ground and splashes outward in all directions. An airplane that encounters a downburst when close to the ground does not have time to compensate for the sudden downdraft, and control is lost. The falling air also produces wind shear. Normally, this term describes sharp changes in wind speed and direction over a short distance. Heavy rain often accompanies a downburst, causing two further hazards for aircraft. First, large amounts of water can be ingested into jet engines, interfering with their operation. Second, thin sheets of water can give the airplane a "ripply" surface, reducing its lift.

A cold bubble of upper air is thought to form when moisture in falling precipitation happens to evaporate. Evaporation is a cooling process, as you easily can feel when water evaporates from your skin. Near the evaporating moisture the surrounding air likewise cools quickly and begins its rapid descent. Recent studies have shown that downbursts are quite common. Nearly two hundred were detected in just one year of observation. A major research effort on downbursts has been mounted at Denver's Stapleton Airport. Rapid progress is being made at understanding and detecting downbursts, a major hazard for aircraft.

69. What causes tornadoes and hurricanes?

Tornadoes are the most violent storms that nature produces. When they occur over water as waterspouts they look like white funnels of water pouring from the clouds.

The formation of tornadoes is not well understood. In the Midwest and Great Plains states, they are sometimes triggered by warm, moist air from the south. Tornadoes are usually associated with intense thunderstorms. Somehow, turbulent rotation within a storm cloud greatly intensifies and reaches toward the ground. Once begun, the whirlpool of air strengthens into a destructive vacuum sweeper.

Tornadoes have wind speeds between 100 and 500 miles per hour. They usually have a damage width of a few hundred feet, a path length of a few miles, and a duration of several minutes. However, tornadoes are very unpredictable. A Kansas funnel in 1990 was ¾ miles wide and lasted 2½ hours before dissipating. Tornadoes generally travel from the southwest toward the northeast. For protection, the National Weather Service recommends that people move to the interior of homes or buildings. Small rooms such as closets and bathrooms offer the best protection from flying debris or roof collapse. In some instances people and entire houses have been picked up by tornadoes and then set down again, almost undisturbed. In Old Testament times, Elijah was taken up to heaven by a supernatural whirlwind (2 Kings 2).

Like tornadoes, hurricanes also go by several names: cyclones in the Indian Ocean, typhoons in the western Pacific, and Willy-willies in Australia. Cyclone is also a general name given to any low pressure region. Hurricanes are giant low pressure, circulating storms that can last for several days. Hurricanes begin over warm tropical waters; their energy source is the condensation of water vapor to liquid raindrops. This vapor-to-liquid transition produces large amounts of energy called latent heat. A mature hurricane moves millions of tons of air at about 200 miles per hour. It may produce 16 trillion watts of power, the equivalent output of 16,000 large nuclear power plants. When a hurricane moves over land its water energy source "dries up" and it soon dissipates.

Although many tropical storms develop each year, few reach the hurricane definition, which is swirling winds of at least 74 miles per hour. The worst weather disaster in United States history was a Galveston, Texas, hurricane on September 9, 1900. This was before weather satellites, and there was very little storm warning. Some six thousand people died in the sudden raging wind and flood. Most deaths were due to storm surge, a mounting up of water and waves along the coastline. Hurricane danger is especially great in places like the Bay of Bengal, west of India. Storms sometimes enter this shallow, restricted region. The resulting high storm surge has killed hundreds of thousands of people in low-lying areas. In November 1913, an unusual storm with hurricane-force winds developed over the Great Lakes, the world's busiest waterway. The storm originated over Canada's Hudson Bay and then raged southward. Radios were scarce in those days. Nineteen ships sank, twenty more were driven into rocks, and 250 sailors drowned in this tragic and rare Midwest storm.

70. What is El Niño?

This Spanish term means "the male child." It refers to an unusual ocean current that begins to flow eastward in the Pacific Ocean about every two to six years. Moving along the equator, the water flows opposite the usual ocean current and brings warm water to Peru along the west coast of South America. This unpredictable ocean current often occurs around Christmas, and the Spanish name thus refers to the Christ child.

El Niño affects worldwide weather patterns in a harmful way. These are some of its far-reaching consequences:

Sea level rises a foot along the South American coast, pushed higher by the counter current. The coastal water temperature increases by as much as 15°F. Commercial fishing is upset, spelling temporary economic disaster for Peru and Ecuador.

Unstable air unleashes torrential rain and severe windstorms in other South American countries.

During the 1976–77 El Niño, the entire bird population of Christmas Island in the South Pacific disappeared. The 17 million birds apparently migrated elsewhere because of temperature changes in the ocean.

In 1982, El Niño is thought to have caused many worldwide problems: droughts and wildfires in Australia, a rare hurricane in Hawaii, flooding in Louisiana, and altogether 1500 related fatalities and $8 billion in damage.

During two weeks in early 1992, El Niño dropped more rain on southern California than is usually received in a year.

The reversed ocean current may be the single most important disruptive influence on world weather patterns. The reasons for El Niño are not known, and therefore its next occurrence cannot be predicted accurately. Historical records show that El Niño has probably been occurring for centuries. The ocean current and its influence remind us that natural events are related in many complex ways. A change in one component of the environment may have global results. El Niño shows us how fragile is our existence on the earth. When nature "misbehaves" slightly, disaster may quickly follow.

71. Does God send deadly storms?

This question goes back to the time of Job. Severe weather had killed his sheep, his servants, and all ten of his children. As Job suffered further physical affliction, he asked, "Why is there evil in the world?" His friends Eliphaz, Bildad, Zophar, and Elihu were not of great help in answering the question. Finally, after much dialogue, God spoke directly to Job out of a whirlwind (Job 38–41). The

Lord made clear that his control over nature and his ultimate purposes are to be trusted. We simply cannot understand all the details of the divine plan.

Storms, floods, and earthquakes are indeed a part of the present world. We sometimes call them "natural disasters," but they are not a surprise to God. This question reveals a serious flaw in secular environmentalism, which sees nature as inherently good and normal. The biblical view goes back to the creation for its "good." Today there are results of the curse operating in nature. The pain and sorrow of this present world, along with severe weather, demonstrate an imperfect world. But a new world is coming, as promised in Revelation 21. Job's testimony demonstrates how trials should bring us closer to God instead of driving us away: "The LORD gave and the LORD has taken away; blessed be the name of the LORD" (Job 1:21 KJV).

Some have concluded that suffering occurs because it is beyond God's control. This is incorrect. God has indeed established certain laws and principles that govern nature, but he remains sovereign over these laws. Psalm 148:8 declares that storms "do his bidding." Concerning Jonah, it was the Lord who "sent out a great wind into the sea, and there was a mighty tempest in the sea" (Jonah 1:4 KJV). Old Testament writers did not hesitate to attribute the forces of nature to God: "thy waterspouts . . . thy waves . . . thy billows" (Ps. 42:7 KJV). Today God's creation is too often depersonalized. He is in control, and has his reasons for all kinds of weather, both fair and stormy.

72. Is it okay to pray for rain?

This question arises from two different concerns. First, is it proper to pray for a change in weather patterns? After all they have already been established by the wise Creator. Second, do we have the right to ask for rain we may need, but which may be a problem for others (it may ruin a picnic or flood a neighborhood)?

God certainly does want our prayers, including petitions. If they are made in sincerity and in faith, he will answer them in the best possible way, including a rain shower. The first concern hints that God is somehow "locked in" by immutable natural laws that he established long ago. However, we must be careful not to put these natural laws on an equal level with God, which would be a form of pantheism. Instead gravity, weather cycles, and planet motions are established to serve humanity. The Lord can alter them at his pleasure. It sounds very complicated to us for a stormy sea to be quickly calmed (Mark 4:39). However, it takes no special effort on the Lord's part; only his word.

In Scripture are examples of prayer for an end to drought, which were answered. "[Elijah] prayed earnestly that it would not rain, and it did not rain on the land for three and a half years. Again he prayed, and the heavens gave rain, and the earth produced its crops" (James 5:17, 18). In 1 Kings 17:1 we learn that not even dew formed during this entire time of drought. Also, we are invited to make weather requests: "Ask the LORD for rain in the springtime; it is the LORD who makes the storm clouds" (Zech. 10:1).

The second concern is that rain may be a blessing for one person but a problem for another. However, this reasoning could wrongly be used to rule out any prayer whatsoever. As does the first concern, this one also hints that God does not have complete control, and that when one person is rewarded others must somehow suffer. However, God is entirely able to do what is best for all of the parties involved. He is sovereign. It is impossible to second-guess what all the indirect consequences of a prayer will be, and we shouldn't try.

One of the givens of the prayer life is that in most matters about which we pray, we don't know what the will of God is. But we have the wonderful privilege of making our requests known to God (Phil. 4:6), whatever those requests may be. We make our petition and leave the disposing of it to our all-wise and loving Father, who knows just exactly what is right. This applies to the matter of praying for rain. This is a comfortable way to live, as the old-timers used to say. Unselfish prayer offered in humility and faith, whether for rain or health, is proper at all times.

PART 4

Past Weather

73. What was the pre-Flood climate like?

The world's early climate was probably very different from today. Uniform and pleasant, some might have called it a tropical paradise. Perhaps there were no deserts or cold polar regions, and no strong winds or severe storms. Long periods of twilight would have occurred in the polar regions just as they still do today, due to the tilt of the earth. However there is evidence that there were no frozen wastelands. Instead, an invisible vapor canopy may have provided a pleasant greenhouse effect for the earth (Q. 93). Solar energy would penetrate the vapor canopy as through greenhouse windows, and warm the entire earth's surface to an ideal temperature. We recall that in the Garden of Eden, Adam and Eve were originally unclothed yet comfortable. This early climate also provided perfect conditions for the growth of plants and animals. It was a time of many dinosaurs that roamed a world that was tropical in climate. The concept that early people struggled with a violent animal world or a severe climate is not taught in the Bible.

The abundance of fossil fuels—coal, oil, and gas—testifies to a planet that once was warmer. These fuels consist largely of tropical vegetation that has been buried and compressed. Today these resources are often found in regions of harsh climate: the deserts of the Middle East, Alaska's North Slope, and under the oceans. The conclusion is that many of these locations at one time supported lush vegetation. As a further example, one of the earth's largest underwater land formations was once covered with forest. This submerged plateau is located in the Antarctic, one-half mile deep in the Indian Ocean. Its vast size is four times that of Texas, or about one million square miles. Deep-sea drilling has brought up fossil wood, leaves from beech trees, and soil samples. This fertile land mass and many others were later inundated and sunk by the Genesis Flood waters.

74. What was the vapor canopy?

This layer of invisible water vapor may have surrounded the earth from the time of creation until the Genesis Flood. There is debate among creationists about the existence of such a canopy; at this time it is one possible model for explaining the past. It could have warmed the early earth for centuries, then collapsed as a major source of the Flood waters. The early canopy would not have been visible in the sky. Its water content was in the form of water vapor or humidity. The sun, moon, and stars would shine through the canopy, somewhat diminished in brightness.

Studies show that a pre-Flood temperature inversion could have supported a vapor canopy. Other models have considered the canopy to consist of ice or liquid water. Despite many unanswered questions, evidences for an early earth canopy are numerous:

1. The initial separation of water above and below the firmament (expanse, sky, atmosphere) may refer to the origin of a canopy (Gen. 1:6, 7).

2. Genesis 2:5, 6 hints that the early earth did not experience rain (Q. 75). A canopy would have greatly affected weather patterns, very likely limiting precipitation.

3. During the great Flood, the windows or floodgates of heaven were opened for forty days and nights (Gen. 7:11, 12). The present-day atmosphere could not hold enough moisture for such a lengthy deluge. Thus, the sky was likely different in the past in a way that could have included a vapor canopy.

4. Many ancient cultures have traditions telling of a canopy-type phenomenon above the earth. The Babylonian creation account, *Enuma Elish,* refers to *Mummu*, representing fog, clouds, or mist. Perhaps this was a canopy description.

5. There is abundant evidence of a warmer earth in the past (Q. 73). We have found fossils of palm trees in Alaska, crocodiles in New Jersey, and petrified wood in the desert. A canopy would have warmed the entire earth like a blanket. Seasonal cold and heat are first mentioned after the Flood, in Genesis 8:22.

6. A canopy would have increased atmospheric pressure, perhaps doubling its present value. This may explain the mystery of the flying reptiles, such as the pteranodon with its seventy-foot wingspan. In today's world it is doubtful that such creatures could fly. With higher air pressure in pre-Flood times, however, they may have been buoyed up by the increased air pressure. These magnificent creatures were certainly created and designed to fly above the early earth.

7. The canopy may hold the key to understanding longevity in Old Testament chronologies, and also giantism of plants and animals in the fossil record. Increased air pressure can be very beneficial to health, as evidenced by experiments with hyperbaric, or high air-pressure surgery. The healing process is sometimes found

to accelerate when the patient is placed under higher than normal air pressure. The vapor canopy may also have increased the human lifespan by absorbing harmful radiation. The aging process is still not well understood today, so it is difficult to speculate on the beneficial effects of a pre-Flood canopy.

A biblical hint that a type of canopy may be restored during millennial times is in Psalm 148:4, 6, which describes waters that will be above the heavens for ever and ever. Isaiah also mentions future weather protection above Mount Zion (Isa. 4:5, 6). See also Q. 100.

75. Was there rain before the Flood?

The rain falling on the roof of the ark may have been the very first rainstorm on earth. Several lines of biblical evidence of this exist:

1. No mention is made of rain on the earth until the Flood (Gen. 7:4, 12). The original earth and Garden of Eden were watered by streams, rivers, and mist instead of by rain (Gen. 2:5, 6, 10). These sources may have been replenished from groundwater. Humidity and mist are still effective today in watering plants. Part of Adam's responsibility in the garden may have been to provide irrigation for the vegetation (Gen. 2:15).

2. The vapor canopy that may have existed prior to the Flood would have greatly affected climate. It could have ruled out rain showers. With a uniform temperature over the entire earth, there would not have been the significant high and low pressure regions that produce storms today. From the moment the canopy collapsed, rain would then become an everyday experience.

3. The rainbow represents a special covenant or promise of protection from another worldwide flood. The rainbow's appearance to Noah may have been its first occur-

rence in the sky (Gen. 9:8–17). Typical raindrops of sufficient size to cause a rainbow require atmospheric instability. Prior to the Flood, weather conditions were probably very stable.

If the earth did not experience rain before the Flood, then Noah's ark-building project must have appeared especially foolish to his critics. Likewise, the faith of Noah described in Hebrews 11:7 was especially strong. Noah was warned about "things not yet seen," which is perhaps a further indication that rain was not part of humanity's early experience on earth. Even with this accumulated evidence, a final authoritative answer to the question of pre-Flood rain is not known.

76. Where did the Genesis Flood waters come from?

Critics often claim that a worldwide flood is completely impossible. After all, they explain, the atmosphere today only holds an average of two inches of water. Of the earth's total water, only 0.0001 percent is held in the atmosphere. However, in spite of the skeptics, Genesis 7:11, 12 gives two adequate sources of water for the universal flood: "On that day all the springs of the great deep burst forth, and the floodgates of the heavens were opened" (Gen. 7:11).

First, fountains or springs of ground water were released on a worldwide scale. Still today, vast amounts of water are stored below ground. About thirty times more fresh water exists underground than is visible in all the surface lakes and rivers. Some of this ground water is under great pressure and heat, as demonstrated by springs and geysers such as Old Faithful in Yellowstone National Park. Even more dramatic, the volcanic explosion of Mount St. Helens in 1980 was set off by a steam explosion from superheated ground water. This water was heated by underground magma, and then released during an earthquake and avalanche on the mountain. This powerful volcano provides a small glimpse of the multiple explosive uprisings of water that may have accompanied the Flood of Noah's day.

A second source of flood waters was the opening of the floodgates or windows of heaven. If there was a greater amount of moisture in the early atmosphere, perhaps in the form of a *vapor canopy* (Q. 74), then the collapse of this water vapor upon the earth over several weeks as torrents of rain would well be described as the opening of the floodgates of heaven. Raindrop formation may have been triggered by volcanic dust explosions. However, God may have had his own methods for beginning the Flood.

77. What was weather like during the Flood?

It is surely impossible for us to visualize accurately the details of the Genesis Flood. Even the time scale completely surpasses our experience with storms:

Forty days of rain (Gen. 7:11–23).
Water covers the earth for 150 days, including the forty days of rain (Gen. 7:24–8:4).
Forty days after the mountaintops first become visible, the raven and first dove are sent forth (Gen. 8:6–9).
Noah and his family finally leave the ark 371 days after entering it (Gen. 8:14).

Air temperature during the Flood would be mild. As water vapor changes to liquid drops, latent heat is given off. Scripture also does not picture a raging storm during the Flood, only a steady mounting up of waters from the floodgates of heaven and from the great deep. Perhaps we incorrectly picture the ark as being buffeted by a stormy sea under a sky filled with violent lightning. Instead, there may have been a continuous, almost gentle downpour.

No wind is mentioned until the end of the Flood. At the conclusion of the Flood, Genesis 8:1 says that the waters receded as wind blew across the earth. This was a refreshing, cleansing wind, opening up the post-Flood world. The word *recede* implies a gradual subsiding of the waters off the land and into the sea. Thus, even the end of the Flood may not have been a violent period. This "gentle" description is countered by the obvious, large-scale twisting and erosion of the earth's upper surface rocks. Also, places like the Grand Canyon give evidence of catastrophic change. Conditions may have been very diferent above and below the water. A definite answer to the question is uncertain.

78. What happened to the Flood waters?

A clue to the answer may be found in Psalm 104:7–9, a passage that describes the earth's creation. The Genesis Flood is also in view in this reference, since it mentions the rainbow promise of no more worldwide floods:

> But at your rebuke the waters fled,
> at the sound of your thunder they took to flight;
> they flowed over the mountains,
> they went down into the valleys,
> to the place you assigned for them.
> You set a boundary they cannot cross;
> never again will they cover the earth.

The Flood waters are said to have flowed over the mountains and down to the valleys. Today, the greatest valleys on the entire earth are the submerged ocean basins, some as deep as seven miles beneath the sea surface. And great mountains exist on the ocean floor, the peaks of the highest ones forming islands. As the Flood concluded, great tectonic activity evidently pushed the continental land upward and the ocean basins downward. Of course, every physical change on earth occurs by the hand of the Lord, using either natural or supernatural means. The conclusion is that the Genesis Flood waters are still present on the earth, as close as the nearest shoreline. Great heights such as Mount Everest were formed after the Flood, having been pushed upward as the ocean valleys were lowered. Various sedimentary rocks, marine fossils, and shells are frequently found on today's mountain tops. The highest peaks also appear to be relatively youthful and only recently eroded. Thus most are post-Flood formations. Of course, some mountains also existed prior to the Flood, according to Genesis 7:20, although their heights are unknown. Perhaps these mountains were located in a lowland area so that their actual height above sea level was not great. Either the waters rose high enough to cover these pre-Flood hills and mountains, or they may have been worn downward by the Flood waters. Whichever the case, the testimony from Scripture and from the earth declares a worldwide flood of judgment in the days of Noah.

79. Was there an ice age?

There is strong evidence that ice once stretched far beyond its present boundaries of the Arctic and Antarctic regions. The Ice Age resulted in deformed rock layers, hills of rock debris that were transported by ice, and flat areas where large amounts of melt-water flowed outward. Across upper North America, including the Midwest, a vast continental ice sheet reached a thickness of thousands of feet. Try to visualize this ice mass, standing as high as small airplanes fly. The land underneath was actually depressed by the great weight of overlying ice.

Many creationists believe that this Ice Age occurred directly following the Genesis Flood, since the ice evidence is still present today. Until the vegetation fully recovered following the Flood's devastation, the earth's climate was cooler than at present. Plants and trees are very effective in absorbing the earth's sunlight energy and warming the earth. Other cooling factors may have been increased volcanic dust and cloudiness, or reduced atmospheric carbon dioxide. The post-Flood, colder period of time probably lasted for several centuries. It was a difficult time for many creatures, including the tropical dinosaurs that had been protected on the ark. There may be hints of a cooler earth in the Book of Job. It was written about four thousand years ago, centuries after the Flood. It mentions snow (6:16; 24:19; 37:6; 38:22), ice (37:10), and cold (38:29, 30).

Modern geologists have continually revised their estimates of the Ice Age era. Many of them, both secular and creationist, now date the Ice Age as only a few thousand years ago. Some geologists speak of multiple ice ages because of overlapping layers of glacial evidence. Most creationists favor a single ice age, with the extent of ice advancing and receding several times as the temperature and precipitation fluctuated. This resulted in later ice layers and debris being deposited over previous layers. A colder climate following the Genesis Flood is consistent with the scientific evidence. In contrast, secular science has no clear explanation for an ice age. The usual claims of a cooler sun, change in earth's orbit, or a large meteorite collision with earth all have serious scientific difficulties.

80. What froze the woolly mammoths?

Woolly mammoths were once widespread in northern latitudes, especially across Alaska and Siberia. They were slightly smaller than today's African elephants, with long curving tusks and thick coats of wool-like hair. They had broad feet, convenient for grazing in swamps and grasslands. Their smaller cousins, the mastodons, lived farther south from the Midwest to Florida.

Mammoth remains, estimated in the millions, are found today frozen in the northern permafrost, or permanently frozen ground, along with remains of sabertooth tigers, giant ground sloths, short-faced bears, and native camels.

Many theories deal with the extinction of the mammoths. As with the dinosaurs, secular science has no generally accepted solution to the puzzle. However, a likely creationist perspective on the mammoths is slowly emerging. They certainly lived in pre-Flood history and thus were part of the animal collection on the ark. Then in post-Flood times they rapidly multiplied and spread throughout the northern regions of the earth. Question 79 explains the arrival of the Ice Age following the Flood. In the initial stages, the far north would not have experienced severe cold. Instead, the Arctic Ocean, bordering Alaska and Siberia, held relatively warm water. This brought abundant rain and vegetation to the mammoths' region, and probably resulted in an animal population explosion. In the later stages of the Ice Age, the climate in the far north deteriorated. Vegetation decreased and water sources dried up or froze solid. Evidence includes a frozen baby mammoth, found in Siberia in 1977, showing severe signs of starvation. The earth finally warmed because of reforestation, and the glacial ice began to melt. Great rivers flowed northward in Siberia, and large-scale deposition and erosion of soil occurred. Many woolly mammoth remains were buried in muck and gravel deposits at this time. The northern reaches gradually became frozen and have remained so ever since. Geologists place mammoth extinction in the cold Pleistocene epoch, about 10,000 years ago. Creationists would generally cut this time in half to about 5,000 years ago.

Some researchers have suggested that a supercold catastrophe killed the woolly mammoths, with temperatures suddenly dropping down to –150°F. This is based on the rare occurrence of frozen mammoths with undigested, unspoiled food in their mouths and stomachs. More likely, a few mammoths were especially well preserved by rapid burial. The temperature decrease was probably gradual. If temperatures had fallen drastically, all life would have frozen solid, and even birds would have dropped from the sky. Actually, birds and most other small animals appear to have migrated out of the mammoths' region. Farther south, several mammoths have been found with embedded spears and arrow points. Thus, hunters were also on the scene several centuries following the Flood. They had crossed into North America over the exposed Bering land bridge connecting Asia with Alaska.

81. Did a climate change kill the dinosaurs?

The answer is probably yes, although there is an alternative explanation to that of secular science. A large number of theories have been suggested for the demise of the dinosaur:

A comet or meteorite hit the earth, raising a dust cloud which shielded the sun. Plants stopped growing, and the food chain broke down.

Many volcanoes erupted, similarly filling the atmosphere with dust.

The sun lessened in its energy output for a time. Winter seasons became severe.

New kinds of plants evolved which poisoned the dinosaurs.

The earth's tilt was changed due to a collision with a large meteorite or planetesimal. This brought severe seasonal changes.

Each of these ideas lacks convincing evidence.

Representative kinds of dinosaurs were carried on the ark. Most remaining dinosaurs perished and some were fossilized. Many creationists believe that a temporary, colder climate then followed the Genesis Flood (Q. 79). During this time, post-Flood dinosaurs, along with many other plants and animals, had difficulty adjusting to the colder climate. Thus a temporary, cold climate probably killed the remaining dinosaurs several thousand years ago.

82. What is the Cretaceous-Tertiary extinction?

This is allegedly a time of severe weather conditions on earth 65 million years ago. The dinosaurs are commonly thought to have died out suddenly during this catastrophe. Creationists take issue with details of the extinction story, but it is interesting nonetheless. It begins with a collision between the earth and a large asteroid or comet. Another version starts with multiple volcanoes. Whatever the case, there was a mighty explosion. As a result, firestorms swept through the earth's forests. Massive amounts of dust and soot spread upward into the atmosphere. This resulted in months of darkness and cold, a prehistoric nuclear winter (Q. 96). Rain and snow were laced with strong, poisonous acids from pollutants in the air. It is thought that many animals, including dinosaurs, were either killed directly by the explosion or by a breakdown of the food chain. From the fossil record it is estimated that 90 percent of all plant and animal forms died during this period.

Scientists continue to debate the evidence for a collision or volcanoes and to search worldwide for asteroid impact sites. Mexico's Yucatan Peninsula has been mentioned as a possibility. There have indeed been many collisions in the past, both large and small. However, to connect a specific impact with the death of the dinosaurs is guesswork. Publicity concerning an *iridium* concentration at the Cretaceaus-Tertiary boundary is based on the fact that this particular element is quite common in meteorites. However, the iridium layer is not as sharply defined as sometimes reported. A second problem is that there is not really one specific point in the fossil record where all the dinosaurs become extinct. Some appear to have lived long into the Tertiary period following the Cretaceaus period.

The creationist alternative to the speculative Cretaceous-Tertiary catastrophe story is that the Genesis Flood was a time of mass death and burial. Probably many volcanoes were involved in the breaking up of the "springs of the great deep" (Gen. 7:11). Volcanoes may be the source of iridium concentrations in the rock layers, as recognized by many researchers. The Flood did not result in plant and animal extinction, but the world's air-breathing life largely started afresh from the creatures protected on the ark. The post-Flood Ice Age later resulted in additional fossil accumulation (Q. 80).

83. Was Greenland once green?

This huge island is mostly white today, not green. Its 840,000 square miles are 84 percent covered with ice, averaging 1,000 feet in thickness. Greenland's small population of 54,000 lives along the coast, which does indeed turn green during the short summer.

Some citizens trace their ancestry back to Norsemen from Scandinavia who first arrived in A.D. 986. These pioneers apparently named the island Greenland to attract more settlers. Some might call it an early example of "false advertising." For a while, oats and barley could be grown in Greenland and also Iceland. Early citizens were able to bury their dead in soil which later became permanently frozen. After several centuries, the climate of Greenland gradually worsened during the Little Ice Age period (Q. 86), which lasted for several centuries. Farming conditions deteriorated and the Norse settlers died out or moved away during this time of extreme cold. Years later, new pioneers arrived on the vast northern island. Today, Greenland has been granted home rule by its previous owner, Denmark. The main industry is fishing, since the land does not offer a substantial livelihood. Temperatures average 29°F on the coast and much colder in the interior.

If the question of Greenland's climate is taken back further into history, the answer is completely different. Abundant evidence shows that the entire world was once warm and tropical. Fossil evidence from the interior of Greenland includes palm and fern leaves. Thus Greenland was indeed green at an earlier time. Possibly Greenland did not exist in the pre-Flood world. Instead, the continent may have formed during the Flood's vast tectonic activity. In this case the vegetation may have actually flourished during the early stages of the Ice Age, while the surrounding water was still relatively warm.

84. What was the climate of Jesus' time?

The climate of Palestine has not changed appreciably since New Testament times. It is a warm, semidesert region. Crops still grow today in the same locations mentioned in Scripture: barley in Judea, olives in Galilee. Abundant limestone results in a porous ground surface with few streams. There are seasonal rainy and dry seasons. Scripture describes the early and latter or autumn and spring rains (Deut. 11:14; Hos. 6:3; James 5:7). The early rains come in the autumn, the usual time of planting. This rain softens the sun-baked fields so they can once again be plowed. Planting continues through January. The latter or spring rain occurs during March and April, just before the time of harvest. The Scripture references speak of God's faithfulness in bringing these important rains.

Recent studies have shown that the climate of Jerusalem is unique in the entire world. This area receives the best possible combination of rain and sunshine. It experiences an average of nearly ten hours of sunshine each day of the year. This value is especially large, since Jerusalem is located at a high altitude of 2500 feet above sea level. The city is bordered by the Judean Desert, which produces very few clouds. Many cities receive more sunshine, but they all lack precipitation. Rainfall in Jerusalem totals 20 inches per year, far more moisture than other desert or semi-arid areas receive. The Jerusalem region is indeed a land of "milk and honey" (Exod. 3:8) compared with the surrounding Near Eastern deserts. The conclusion is that the climate of the Jerusalem area was beautiful in Jesus' day just as it still is today. Jerusalem is spiritually significant and also unique in climate (see reference).

85. Has God controlled history with weather?

Job 37:13 declares that God sometimes brings storms to punish people. This brings to mind the experiences of both Noah and Jonah; storms played a major part in their lives.

Scripture is filled with many weather-related events used by God to direct people and nations.

In more recent centuries the weather continues to show the providence of God in history, especially in military events. Four centuries ago, Philip II of Spain set out to rule the world. His goal was to reunite all of Christianity under the Catholic faith, and also under Spanish control. Philip was a heartless leader who had earlier supported the harsh measures of the Spanish Inquisition. He did not get along well with either Britain or Holland. When the Netherlands declared their independence from Spain in 1588, Philip sailed north with an "invincible" Armada of 130 warships and 30,000 men. Then, while far from home, bad weather hit the fleet. With no port available the ships were forced to anchor at sea. The smaller British fleet under Sir Francis Drake then harassed the Spanish by releasing flaming "tar ships" among the anchored fleet. The Spanish Armada fled north through the English Channel and around the coast of Scotland, where continued fierce storms pushed the fleet onto the rocky shores. As the history books used to say, "God blew and they were scattered." Only half of the original Spanish fleet finally limped back home. The great Spanish sea power had been diminished by the weather, instead of by war. Perhaps a worldwide dictatorship was averted by this defeat of the Spanish Armada.

In a famous painting, General George Washington is shown crossing the Delaware River on Christmas night 1776. With snow falling, his forces were able to surprise and defeat the enemy. This victory fanned the revolution's fading embers. Weather continued to play an important role in America's Revolutionary War.

In June 1940, 400,000 British and French soldiers were trapped on the north coast of France at Dunkirk. The Nazi forces were only ten miles away and could have devastated the cornered Allied troops. However, at this time thick fog and clouds settled over the beaches. The English Channel also became unusually calm, and many small boats from England were able to ferry soldiers across to safety. The protective weather pattern lasted nine days while the men were evacuated. The Allied army was thus preserved to fight another day. Winston Churchill called the evacuation of Dunkirk a "miracle of deliverance."

These examples show how history has indeed been shaped by the weather. In turn, the weather itself is directed by the one who established it. A recognition of this truth may cause us to think twice before complaining about the weather.

86. What was the Little Ice Age?

The average earth temperature dropped slightly during the centuries between A.D. 1400 and 1850. The coldest part of this period, 1645–1715, is called the Little Ice Age. This time of cold was accompanied by many worldwide effects:

The Arctic ice expanded, isolating Iceland and Greenland from shipping.

Tree lines on the mountains of Europe dropped to lower levels. That is, trees did not grow as well in the cold climate.

The production of grapes in England ceased. The Thames River flowing through London froze during the winter seasons, unlike today. Citizens held "frost fairs" out on the ice.

The year 1816, toward the end of the cold period, was known as the year without a summer. Weather was not helped by a large volcanic eruption in the Indian Ocean a year earlier. Ice and snow appeared in the eastern United States every month during 1816; people wore overcoats in July. Pioneers labeled the year "eighteen hundred and froze to death." The potato famine in Ireland also occurred during this time, accompanied by a typhus epidemic that killed 65,000 people. The cause of the colder period of weather is not certain. One possibility is an unusual lessening of sun activity, including sunspots, that occurred during this time. Sunspots—magnetic storms on the solar surface—show up as dark patches. The number of sunspots may be linked to the earth's climate. Others believe that volcanic dust accumulated in the upper atmosphere and reflected the sun's rays back into space, thus cooling the earth for a time.

87. What caused the dust bowl of the 1930s?

The Great Plains states include Kansas, Oklahoma, and surrounding states. This region receives scanty rainfall, but the natural prairie vegetation of thick grass and few trees once covered the rich soil. In the early 1930s, large numbers of settlers moved into the area and plowed the prairie. The dust problem began with the intense farming of marginal land without consideration of the consequences. Deprived of its cover of tough grass, the soil became vulnerable to wind erosion. Several years of dry weather added to the problem for the weary farmers. There was a popular belief that "rain would follow the plow," but rain did not come. When the wind blew, great clouds of dust and soil filled the air. The dust was like rolling black smoke and sometimes suffocated cattle in its path. It covered fences and almost buried farm houses. In places large depressions over one hundred feet deep formed in farm fields, called dust bowls. Some of these eroded areas have never recovered, even after sixty years. We are reminded of the Old Testament rules of caring for farm land, such as the sabbatical rest (Exod. 23:10–13).

88. Has the sun's heat changed in the past?

On an evolutionary time scale, scientists are forced to conclude that the sun is now 30–40 percent warmer than in the distant past. This is based on natural changes in an aging sun during a five-billion-year lifetime. A cooler sun would have greatly affected the early earth's climate. Glaciers would completely cover the land, the oceans would be frozen solid, and life on earth would be impossible. This multi-billion-year scenario is not connected with the Ice Age, which occurred just a few thousand years ago.

Evidence for a cold, early earth is completely lacking. Instead, the oceans have apparently remained liquid during all of earth's history, even on a long time scale. Fossilized marine microbes show that the oceans have always remained warm. Some scientists, locked into the long time scale, propose that an extensive atmosphere of carbon dioxide on the early earth might have kept it warm while the evolving sun slowly warmed up. However, this situation could easily have led to a runaway greenhouse effect (Q. 93) and severe 900°F conditions, as on planet Venus. To avoid this problem, these scientists further suggest that blue-green algae evolved just in time to use up the excess carbon dioxide. This allowed the earth to maintain a balanced temperature as the sun gradually warmed. In this view, the comfortable temperature of the earth is said to depend on an evolutionary accident. In contrast, the recent-creation view maintains that there has not been unlimited time during which the sun evolved. Instead, the sun shines today just as it did when created on the fourth day of the creation week. This is a simple, adequate, and satisfying explanation of why the past and present climate on earth is relatively stable (see reference).

89. Has our atmosphere evolved over time?

In the creationist view the atmosphere was fully functioning by the third day, when plants appeared; on the fifth day, birds flew and fish swam the seas (Gen. 1:11, 20). Thus there is no biblical reason for a slowly evolving atmosphere.

In the naturalistic view, however, the earth's atmosphere is required to have changed drastically over time. Air on the early earth is assumed to have come from volcanic gases and from meteorite bombardment. Some scientists believe the primitive earth had gases somewhat similar to present-day Venus and Mars: carbon dioxide (CO_2), nitrogen (N_2), water vapor (H_2O), hydrogen (H_2), carbon monoxide (CO), methane (CH_4), ammonia (NH_3).

Scientists have no general agreement on which of these gases might have dominated the early atmosphere, and, in fact, engage in considerable debate over the early existence of methane and ammonia.

This supposed early atmosphere is called reducing; it would have supplied electrons to substances and chemically reduced them to other forms. This reducing process often separates molecules into smaller components. Thus the early atmosphere would actually have broken down any building blocks of life that were present.

Other speculations on the forming of today's atmosphere are also highly questionable, such as these:

Anaerobic (without oxygen) bacteria came to life in the sea.

> Blue-green algae evolved and began making oxygen by photosynthesis. A half-billion years ago the oxygen level finally reached its present value.

Clearly this speculation is closely connected with the assumed evolution of life. However, geochemical evidence for the assumed early atmosphere is lacking. It appears that evolutionary theory has added unnecessary complications to our understanding of the earth's atmosphere.

Some "old earth" creationists have attempted to find divine direction in the evolution of the earth's atmosphere. They emphasize that blue-green algae came along at just the opportune time, making oxygen in preparation for higher forms of life. They also suggest that early, heavy carbon dioxide concentrations dwindled just in time to avoid runaway greenhouse warming on earth. Thus God directed a slowly changing atmosphere for our eventual benefit. This reasoning is certainly interesting, but these cautions are advised:

1. If the young-age view is correct, then this compromise with the evolutionary time scale is misleading.
2. No general agreement or evidence exists that the atmosphere was once abundant with carbon dioxide, or depleted in oxygen.
3. Secular scientists usually are not impressed by "theistic evolution" arguments. The atmosphere did somehow evolve in step with life, they say, regardless of probabilities against it.
4. In Genesis 1:11, land plants and trees are described as the first life on earth. The compromise view begins with algae in water, and then with land vegetation appearing long afterward, about one billion years later.

The idea of a slow, providential alteration of earth's air in preparation for life sounds appealing at first. However, accommodation to long-age, evolutionary views eventually leads to deeper problems. The refreshing alternative is that the earth's atmosphere was created with maturity, complete with the oxygen that our first parents, Adam and Eve, needed.

90. Is helium in the air an age indicator?

Much available data indicates that the earth may be relatively young or recent in age. The amount of atmospheric helium is one such measurement. Helium atoms are formed in the rocks of the earth as radioactive materials decay. A helium atom is an *alpha* radiation particle that has gained electrons. Current estimates are that more than a million helium atoms are released from each square inch of the earth's surface every second. Thus a fine spray of helium gas continually moves upward from the ground and into the sky. A small amount of this helium gas escapes into space and the rest accumulates in the upper atmosphere. It is found that the current total of atmospheric helium would take "only" about 2 million years to accumulate. This seems to provide an upper limit on the age of the atmosphere. If the rate of helium flow was greater in the past because of greater radioactivity, then the age limit becomes even shorter. Clearly this raises questions about the evolutionary age of the earth and atmosphere being 4.5 billion years. This long age is 225 thousand percent greater than 2 million years. On such a long-age scale, where is all the missing helium? It has been suggested that the solar wind somehow sweeps helium away from the atmosphere, but evidence is lacking (see reference).

No physical measurement can be expected to precisely date the earth. Every age experiment is loaded with assumptions and uncertainties. A large spectrum of earth ages can be obtained, ranging from only hundreds to many billions of years. Many creationists emphasize the younger ages, because these usually involve the least amount of guesswork. The helium data, as one example, promotes the theory of a recent creation. Those who support an ancient earth are challenged to explain the helium data. Likewise, creationists work to give alternative explanations for long-age measurements, such as radiometric dating. In this way creation science serves a valuable purpose in questioning basic scientific assumptions such as great age.

PART 5

Future Weather

91. Is the earth warming?

There is no general agreement on the answer to this question. Many experts believe that global warming is inevitable because of the greenhouse effect (Q. 93). One study predicts a 1° to 5°F temperature rise during the next fifty years. However, it is too early to observe such a climate change. In 1990 the United States' average temperature was 53.9°F. This value was 1.5°F above the long-term average, which may indicate the beginning of a trend or simply a one-year variation. On a global scale, average temperatures are very difficult to measure. During the longer period 1880–1970, it is believed, the earth actually cooled by about 2°F (see reference).

One problem in predicting earth warming is the uncertain impact of many factors. For example, clouds are able to both heat and cool the planet. They cause warming by acting as a blanket, reducing the amount of surface heat radiated into space. Estimates are that clouds warm the earth by about 9°F. In a reverse way clouds also absorb incoming sunlight and reflect it back into space, preventing it from warming the earth. It is thought that this reduces surface temperatures by about 22°F. Thus it appears that clouds have a net cooling effect. A warming earth would produce more clouds, which would then work to stop further warming. This "thermostat" process could prevent any substantial global warming. It is one of the many complex processes that God has built into his creation to provide stability. Nature is resilient and able to "bounce back" from extreme conditions. Whether or not this ability could be upset by a severe greenhouse effect is not known.

Some extreme environmentalists have proposed the Gaia hypothesis, that the earth is in some sense "alive." The name comes from the Greek goddess of the earth; this view fits closely with the New Age religion. Gaia's followers believe that evolutionary change is directed by a spiritual force within the "conscious" earth. One "evidence" is the ability of the earth to maintain a near-constant temperature, just like people. Supporters of Gaia further believe we should return to the pagan rites of the ancients in worshipping the earth. Instead of an advancing new age of enlightenment, Gaia supporters have returned to superstition. In sharp contrast, Christians can honor the Lord through stewardship of his creation.

92. What would a warmer or cooler earth be like?

Regarding the earth's future climate, scientists are found on both sides of the temperature issue. Some believe that global warming has already begun; a few predict the approach of another ice age. This book has taken the cautionary position that no one knows the future climate. Too many interrelated elements are involved. It may well be that these factors will counterbalance each other and maintain our present climate. Nonetheless it is interesting, and somewhat unsettling, to consider some possible results of just a 5°F change in the earth's average temperature:

A warmer earth

1. Sea level could rise 25 feet or more. This would result from melting ice and from thermal expansion of the seawater. Worldwide, coasts would be inundated. Florida would be reduced to half its present size. Other submerged areas would include the Nile Delta, Chesapeake Bay, entire islands, and most of Bangladesh. The inland migration of large groups of people would be politically challenging. Cities such as New Orleans, Bangkok, and Venice would require dikes similar to Holland's.
2. Salt water intrusion into groundwater aquifers would ruin millions of water wells worldwide. The water supply of Miami would be poisoned by brine.
3. Warmer temperatures would increase the number and severity of storms. Hurricanes are fueled by energy extracted from warm seawater. Their range would be extended all over the earth.
4. The sea fishing industry, essential to the livelihood and nourishment of half the world's population, would be greatly disturbed.
5. Cloudiness, precipitation, and vegetation growth would increase in many places. Farming would become productive in northern Canada and Russia. The production of corn and rice would increase worldwide. The great salt flats of our western states might become new "great lakes." Certain other areas would become hot and dry; forest fires would increase in number.

6. Much more electricity would be consumed by air conditioning. This increased power generation could cause a severe energy crisis.

A cooler earth

1. Sea level would drop 10–20 feet or more as snow and ice accumulations increased. The coastal borders of countries would grow as the offshore continental shelves became exposed and dry. The mouth of the Mississippi River might become raging whitewater as water eroded a deep valley down to the lower sea level. Land bridges would once again appear, connecting the continents. It would be a "field day" for archaeologists, searching these long-flooded areas for artifacts. Ocean shipping would come to a halt as ports became stranded far inland. With the flow of oil stopped, highways might become hiking trails for migrating people. Thousands of coastal power plants and water treatment facilities would also shut down.
2. The water table would drop steeply; ponds and streams would dry up. Millions of water wells would go dry.
3. The earth might enter a new ice age, with no way to stop its progress. Canada and the northern United States would once again be covered by thousands of feet of advancing ice. Nations of the north (England, Canada, China, Russia) and south (Australia, South Africa) might be forced to migrate toward the equator. Birds and wild animals would also move with the shifting climate. This scenario almost sounds like a large scale, long-term campout where everyone is pushed into a vast, crowded park. Whether the earth's popu-

lation could tolerate such a situation without total chaos is a problem for sociologists to consider.

4. Climate patterns would be altered. Some of the earth's deserts would expand their boundaries. Famine would sweep across third-world countries, where land would become even more barren and dry. Certain other deserts would bloom because of increased moisture.
5. Increased heating-fuel consumption would strain fuel supplies everywhere.

A slight shift of the earth's temperature, either up or down, would surely have serious consequences for all people. Truly, the earth's near-constant temperature is a great blessing. See Q. 99 for discussion of the rapid climate change during the last days of planet Earth.

93. What is the greenhouse effect?

You may have noticed how the inside of a car warms up on a sunny day if the windows are closed. In summertime the car interior can become dangerously hot. This is an example of greenhouse warming. The window glass lets the sunlight pass through. Once inside, much of the sun energy changes to heat, or infrared radiation, after being absorbed by various objects. The resulting infrared light cannot easily pass back outward through the glass. It is thus trapped inside where it raises the temperature. Greenhouses use the same principle, with the solar heat controlled by fans, vents, and shades.

A scientist named Fourier first explained the greenhouse process in 1827. In the atmosphere three particular gases function especially well as "greenhouse windows." They are water vapor (H_2O), carbon dioxide (CO_2), and ozone (O_3). Like glass windows, these gases are transparent to the sunlight streaming toward earth. Also like glass, the gases trap the resulting infrared heat that is radiated back upward from the ground. The gas molecules increase their vibrational motion as they absorb the infrared heat, and this raises the temperature of the atmosphere. Greenhouse warming of the earth is actually a fortunate process: Outer space is a very cold place. Without greenhouse gases, the entire world would be a frozen wasteland, 60°F colder than at present.

The problem in our day is excess carbon dioxide in the atmosphere. The total amount has increased by 10 percent in the last 50 years. This extra CO_2 largely results from the burning of coal, oil, and gas. The stripping of rain forests also contributes, since trees absorb CO_2 from the air: no trees, no more absorption. Volcanoes also inject great amounts of greenhouse gases into the air. In fact, the human impact on the environment may be negligible compared with natural processes. Whatever the source, the ocean provides somewhat of a counterbalance by absorbing extra carbon dioxide. The effectiveness of the ocean as a "sink" for extra carbon dioxide is a major unknown factor in the greenhouse effect.

Increased CO_2 amounts could eventually raise worldwide temperatures. However, warming of the earth has not yet been verified. Temperature records from the past are quite unreliable. A notorious example of a "runaway greenhouse effect" exists on Venus. There a thick atmosphere of greenhouse gases hold the temperature at a constant, baking 900°F.

Greenhouse warming was once thought advantageous to the earth. The Swedish chemist Arrhenius wrote in 1906: "By the influence of the increasing percentage of carbonic acid [carbon dioxide] in the atmosphere, we may hope to enjoy ages with more equable and better climates, especially as regards the colder regions of the Earth." Today it is realized that greenhouse warming might have negative consequences for the earth (Q. 92).

94. How might rain forest destruction affect our weather?

Most of the tropical rain forests straddle the equator. One-third of the world's total is in the Amazon basin of Brazil. Rain forests usually experience high humidity and daily rainfall. The vast areas of trees have been called the "lungs of the earth." The forests "breathe" by removing large amounts of carbon dioxide from the air and giving off oxygen. It is feared that the loss of rain forests will increase CO_2 levels in the atmosphere and result in global warming. On a more local level, the clearing of rain forest land diminishes the amount of evaporation. As a result, the rainfall decreases and the ground dries out, reducing its ability to support vegetation. Loss of rain forests also reduces the absorption of solar energy. Sunlight reflects off the barren ground instead of being absorbed by trees. This would tend to cause earth cooling, an opposite result from the carbon dioxide increase. If the cooling effect dominates, rainfall and air circulation might decrease worldwide.

Rain forests covered about 30 percent of the earth's land surface in 1950; the "greenbelt" is now at 7 percent and still declining. The trees are being cleared for timber, underground minerals, and farms. In the process, the great diversity of plants and animals in the rain forests is destroyed. Many conservationists consider deforestation the planet's greatest environmental problem. Clearly, the ultimate results are uncertain at this time. We are reminded that trees are created to be pleasant for sight, as well as for food (Gen. 2:9). They have a crucial place in this world.

95. Should Christians be concerned about the environment?

Many people have been turned off to environmental issues by extremists. Environmentalists often seem to have more concern for whales, snail darters, and owls than they do for people. Some of them try to make us feel guilty for being alive and using air.

Christians actually have special insight into this controversy. We know that the present world is temporary. The elements will someday melt with fervent heat (2 Peter 3:10, 12). Still, this is not a license to abuse the earth or to be arrogant toward it. Scripture gives us two specific commands in this regard. The first command, sometimes called the cultural mandate, is found in Genesis 1:28: "Be fruitful and increase in number; fill the earth and subdue it." The first part of the command is being fulfilled, with a current world population of more than five billion people. The subduing of the earth involves rulership. We have been placed in charge of the earth as stewards. The earth does not belong to us; it is the Lord's, including everything in it (Ps. 24:1). We are to care for and manage what has been given to us, and we are responsible for the results.

A second biblical command appears in Genesis 2:15: "The LORD God took the man and put him in the Garden of Eden to work it and take care of it." This was Adam's job description, and it is also a general model for us today. Hebrews 2:8 states that God has put all things under our feet; we are still in charge. To work and care for the garden, or "dress and keep" it (KJV), means to make it orderly, productive, and beautiful. The clear conclusion is that we should indeed be concerned about the deterioration of the environment. Only the Christian has the proper foundation for understanding people's true responsibility to care for the earth.

Consider some of the common objections to environmental concern:

1. "People are more important than things." This is true. There is no profit in gaining the whole world and losing one's own soul (Mark 8:36). Presenting the gospel must have priority. If we love the Lord, however, we will not turn our backs on his artwork in the creation. The creation gives powerful testimony to God's glory; caring for his works is our responsibility. The environment is important because people are important. Their health and well-being depend on it.
2. "Why should we be concerned with a temporary, dying world?" In spite of sin, the creation remains a rich testimony to God's goodness. Stewardship of this silent witness is an act of worship. Our bodies are also temporary, but we care for them. Our minds will someday be made perfect, yet we spend our lives studying and exercising them. The same arguments apply to caring for the present, temporary earth.
3. "Environmentalism is just a form of the social gospel, or even worse, it is New Age pantheism." True environmental care is respect for God's handiwork, not a return to idolatry. Stewardship is the recognition of humanity's high created position, not a pantheistic ruling out of our responsibility before God.
4. "We are told in Genesis 1:28 to subdue the earth." Our dominion should not be harsh; this verse is not a proof text to condone littering. To subdue means to cherish and bring out the full potential of the creation, not trash it.

Environmental issues are a major concern of our day. Christians should do their part to help the earth glorify God. After all, we know the Creator. Of course, we must also spread the gospel message, telling others about the Creator of heaven and earth.

96. What is nuclear winter?

Nuclear winter describes a possible scenario for what would happen to the earth's climate following a large-scale nuclear conflict. If such a grim catastrophe occurred, uncontrolled fires would break out in cities and forests worldwide. This fire storm would send large quantities of carbon soot and smoke high into the atmosphere. The blanket of pollutants would quickly surround the earth and shield the ground from sunlight. As a result, global temperatures would drop and a long, harsh man-made winter would follow. An intense thermal inversion (Q. 12) could occur, with ground temperatures plunging by perhaps 25°F. Permanent dense fog might also develop. Some predict that the human race would be eliminated by this new global ice age. As the ash slowly settled to the earth, the planet would gradually warm again. Perhaps new species would then "evolve" to take over the earth. Notice how an evolutionary bias enters this discussion, so typical of scientific studies.

The concept of nuclear winter, popular during the 1980s, has been questioned by many researchers. They note that the original atmospheric model used to predict severe cold was incomplete. For example, it did not include the modifying effect of the oceans. Actually, the oceans act as a thermostat for changing earth temperatures. If the earth cools, the oceans give off some of their vast reservoir of heat. If the earth warms, the seas absorb additional heat to maintain even temperatures. In addition, events from the 1991 war in Kuwait did not support the nuclear winter concept, contrary to some early predictions. Substantial cooling requires that dust rise at least 12 miles high into the stratosphere. However, the smoke from 625 oil fires remained at lower levels and resulted in only local temperature changes.

Other more complete climate models predict a mild winter or a "nuclear autumn" in the event of a nuclear war. Of course, this newer finding, if reliable, should not cause us to think any more lightly about nuclear weapons. What the different models reveal is that detailed weather predictions are difficult, if not impossible. We can also be sure that humanity will not be annihilated by warfare or nuclear winter. Scripture gives a detailed picture of how the present age will close.

97. What is the future weather on an evolutionary time scale?

The far-future weather forecast from secular science is not optimistic. The following discussion is hypothetical only: Several billion years from now the sun will have burned much of its hydrogen fuel. The sun will begin to expand outward, gradually turning into a "red giant" star. The earth will slowly warm as the sun grows larger and redder in the sky. This heating will be much more severe than the much-publicized greenhouse effect. The weather forecast will continually be hot today and hotter tomorrow. The sun's increasing heat will gradually melt the ice caps, evaporate away the oceans, and bake the rocks. Intense solar radiation will finally extinguish any remaining life on earth. The expanding sun will first swallow up and vaporize planet Mercury, then Venus, and finally Earth. Nothing will be left of the earth and its history except a spray of hot stellar gas.

Clearly, the secular destiny of the earth is quite depressing. In contrast, the Bible explains a far different future for planet Earth. The heavens and earth will indeed be purified by fire: "But the day of the Lord will come like a thief. The heavens will disappear with a roar; the elements will be destroyed by fire, and the earth and everything in it will be laid bare" (2 Peter 3:10). However, this cleansing will come by the hand of God instead of a slowly-expanding sun. It is actually the sun that is temporary and not the earth. Some day the new earth will exist with no more need of the sun, for the Creator himself will provide the light (Rev. 21:22). With the Lord in control the climate will become ideal forever.

98. Do present-day weather changes predict the end of the world?

This topic should be approached with caution. It has long been popular to watch for specific signs of the "last days." Revelation 22:7 says: "Blessed is he who keeps the words of the prophecy in this book." However, such studies should avoid the error of date setting. Too often this practice has led to a loss of Christian testimony. Predictions of the end of history will probably increase as the twentieth century ends and A.D. 2000 arrives. However, the close of the millennium results from a man-made imperfect calendar without any particular divine significance. The Bible warns against specific date setting: "No one knows about that day or hour, not even the angels in heaven nor the Son, but only the Father" (Matt. 24:36; Acts 1:7).

We are also reminded of natural events that will continue throughout the present age:

> Nation will rise against nation, and kingdom against kingdom. There will be famines and earthquakes in various places. All these are the beginning of birth pains (Matt .24:7, 8).

Famine suggests unusually dry weather conditions, which indeed occur often at different locations on the earth. The passage indicates that such events may well accelerate before the end times, since present events are just "the beginning." Question 99 explains more of the specific changes that will someday come in the world's weather. Whether we are near or far from the end times cannot be clearly determined from present weather patterns or any other data from science. As an example, table 2 shows the worldwide fatalities from major storms of several centuries. There is no particular trend of increasing weather violence; there are both quiet centuries and deadly centuries. During times of persecution or disaster, of course, it is natural to assume that the current age is unique. In fact, the doctrine of the Lord's imminent (at any time) return is a strong comfort in times of trouble.

The contemporary warnings about global warming may seem to be a unique, new threat to the earth. However, it must be remembered that this world has experi-

Table 2
A summary of the most deadly floods and storms in history.

Date	Catastrophe	Location	Deaths
1228	Flood	Holland	100,000
1642	Flood	China	300,000
1887	Flood	China	900,000
1911	Flood	China	100,000
1931	Flood	China	3,700,000
1939	Flood	China	200,000
1970	Cyclone	Bangladesh	500,000
1991	Cyclone	Bangladesh	150,000

enced past global change, especially by a worldwide flood and an ice age. Only God knows how much more history is to come on the present earth. In good times and in bad, the Christian can take comfort in the Lord's control of the universe.

Scientists describe a number of events they believe could endanger the future of life on earth:

Radiation exposure because of a depletion of the ozone layer, or the supernova explosion of a nearby star.

Severe drought and heat resulting from the greenhouse effect.

A nuclear winter resulting from atomic warfare.

A collision between the earth and a large comet or asteroid.

Writer Isaac Asimov has summarized one hundred such global disasters in a depressing book titled *A Choice of Catastrophes*. The great variety of unsettling ideas shows that each is only an opinion; none is a certainty. Scientific understanding cannot actually grasp the destiny of the world any better than it can grasp the earth's supernatural origin. Only in Scripture are the Lord's plans for humanity and for the culmination of this age revealed. Through the study of Scripture we can be assured that humanity will not be annihilated either by its own hand or by some natural catastrophe.

Neither science nor speculation can accurately predict the dates of end-time events. However, there is no doubt that history is moving toward a conclusion. Scientists have measured a continuous decline in universal order, expressed as the second law of thermodynamics. This basic law shows that the entire universe is "running down" like a giant clock. The Lord controls the unwinding, and can also step into history whenever he determines.

99. What weather events are predicted for the end times?

One recent summer in the Midwest, dry and windy conditions led to an unusual dust storm. The afternoon sky turned a pale brown color; a person could look directly at the sun as through a filter. People became very uneasy about this unusual, dark sky. Some stopped their cars to look upward in wonder. I was reminded of how much we depend on a normal, familiar appearance in the sky. If conditions are the least bit altered, even by dust, people become upset.

We can only imagine the terror that will follow the drastic changes in the earth as described in the Book of Revelation. At that time the order and dependability of nature will break down worldwide. Revelation 4–18 describes events of the future, a time often called the seven-year tribulation period. Table 3 lists some of the extreme weather occurrences during this period. Many catastrophes, including the greatest earthquake of all time (Rev. 16:18), will accompany these weather changes. The overall impression is of an earth that becomes completely unpredictable. The weather system will shudder and reel with drastic change. This may be the greatest distress of unfortunate people during that fearful time.

Sun and moon darkened

This darkening may result from obscuring clouds or else from an actual extinguishing of solar energy. Either way, the immediate result will be twilight conditions and cold temperatures.

Falling stars

Stars will fall to the earth (Matt. 24:29, 35; Rev. 6:13). This may refer to a failure of actual stars, or to an intense

Figure 8.
A woodcut showing the meteor shower of 1833. This intense shower of meteors or "shooting stars" gave people great fear.

swarm of meteors in the sky. Figure 8 pictures an unusually heavy meteor shower that took place in 1833 (see reference). This picture was first used to illustrate prophetic events in a book titled *Bible Readings for the Home Circle,* published a century ago. The engraving shows an unusually strong display of the Leonid meteor shower that takes place each November. A diary from 1833 reports the event:

Table 3
End-time Events

Some of the unprecedented weather events that accompany the time of tribulation. The events are not listed in strict chronological order.

Reference (Rev.)	Event	Explanation
4:5	thunder, lightning	Scene in heaven; thunder represents the glory of God. Thunder is mentioned 8 times in the Book of Revelation.
6:6–8	famine	Third seal opened, causing crop failure.
6:12	sun darkened	Fifth seal opened, and the earth grows dark.
6:13	stars fall to earth	The entire heavens are changed. The verse may describe an intense meteor storm.
7:1	wind stops	The earth's entire atmospheric circulation system is halted.
8:7	hail and fire, mixed with blood	First trumpet sounds, and deadly precipitation results. "Blood rains" have fallen before, actually rain mixed with red mud. This final rain, however, may well be actual blood.
8:12	a third of sun, moon, stars darkened	Fourth trumpet sounds, upsetting the day-night cycle.
13:13	fire comes down from heaven	Performed by an evil "beast" who is given limited power over the atmosphere.
16:8–9	sun scorches people	This plague contrasts with the former—darkening of the sun.
16:21	100-pound hailstones	Such severe hail is unknown in previous history.

During three hours of its continuance, the day of judgement was believed to be only waiting for sunrise, and, long after the shower had ceased, the morbid and superstitious still were impressed with the idea that the final day was at least only a week ahead. Impromptu meetings for prayer were held in many places, and many other scenes of religious devotion, or terror, or abandonment of worldly affairs, transpired, under the influence of fear occasioned by so sudden and awful a display.

If a simple meteor shower had this profound effect on people, we can imagine the fear of multiplied events in the sky during the last days. For those on the Lord's side, however, there is no cause for alarm. Jeremiah 10:2 reminds us: "Do not . . . be terrified by signs in the sky, though the nations are terrified by them." For the believer, signs in the sky show God's presence and control.

100. What will the weather be like during the millennium?

The millennium is the future thousand-year reign of Christ on earth, which some Christians believe is described in Isaiah 35 and elsewhere. It will be a time of unprecedented peace, joy, and blessing. (Other Christians have different beliefs regarding end-time events.) Isaiah 35:1, 2, 7 declares that desert and wilderness areas will flourish with plant life. This will occur because "water will gush forth in the wilderness and streams in the desert" (Isa. 35:6). Evidently the water table will be altered to allow abundant surface water and springs. Precipitation will also increase in formerly parched areas, perhaps leading to an ideal, Garden of Eden-like setting. Severe weather will end and be replaced by "showers of blessing" (Ezek. 34:26). It will be a time of bountiful harvests, both from trees and from fields (Ezek. 34:27). Jerusalem will be the millennial capital of the world, where the Lord will dwell. This region will experience special weather protection, according to Isaiah 4:5, 6. Perhaps a supernatural canopy will cover the city, providing shade from the sun and shelter from the rain. Whatever the case, the millennial world will be filled with unspeakable blessings. Future weather forecasts will be good every day.

References

Question

25. Shkolnik, A., C. R. Taylor, V. Finch, and A. Borut. "Why do Bedouins wear black robes in hot deserts?" *Nature,* 283 (January 24, 1980) pp. 373–5.
49. Elsom, Derek. "Catch a falling frog." *New Scientist,* 118 (June 2, 1988) pp. 38–40.
53. Allman, William F. "Are no two snowflakes alike?" *Science 83,* 4 (December 1983) p. 24.
62. Viemeister, Peter E. *The Lightning Book.* Garden City, NY: Doubleday and Company, Inc., 1961.
84. Alpert, P. "The Uniqueness of the Jerusalem Climate." *Bulletin of the American Meteorological Society,* 72 (February 1991) pp. 228–31.
88. Waldrop, M. M. "Why is the earth neither too hot nor too cold?" *Science 83,* 4 (August 1983) p. 100.
90. Vardiman, Larry. *The Aging of the Earth's Atmosphere.* El Cajon, CA: Institute for Creation Research, 1990.
91. Lindzen, Richard. "Some coolness concerning global warming." *Bulletin of the American Meteorological Society,* 71 (1990) pp. 288–99. Kellog, William. "Response to skeptics of global warming." *Bulletin of the American Meteorological Society* 72 (1991) pp. 499–511.
99. Hughes, David W. "A mysterious woodcut." *Sky and Telescope,* 74 (September 1987) p. 252.

Suggested Resources

General

DeYoung, Don. *Astronomy and the Bible.* Grand Rapids: Baker Book House, 1989. This book discusses the effects of the moon, sunspots, and the solar wind on our weather.

Lehr, Paul, R. Will Burnett, and Herbert Zim. *Weather.* A Golden Science Guide. New York: Golden Press, 1987. The series of compact Golden Science Guides has helped several generations of youth understand and enjoy nature. The books don't use the word *creation,* but they teach it.

Lockhart, Gary. *The Weather Companion.* New York: John Wiley and Sons, Inc., 1988. This book is listed as a *negative* example. It is highly critical of the Bible, creation, and even of missionaries. It is hoped that *Weather and the Bible* will counter such works.

Weatherwise, a bimonthly magazine from Heldref Publications, 4000 Albemorle St. N.W., Washington, D.C. 20016. Many articles and statistics on meteorology are included.

Technical

Creation Research Society Quarterly. P.O. Box 28473, Kansas City, MO 64118. Research articles are frequently related to the weather and climate.

Institute for Creation Research. P.O. Box 2667, El Cajon, CA 92021-9982. This research center sponsors a number of publications and projects, some of which involve meteorology.

Oard, Michael J. *An Ice Age Caused by the Genesis Flood.* San Diego: Institute for Creation Research, 1990. Meteorologist Oard explains a possible mechanism for the post-Flood Ice Age. Technical details and a large bibliography are included.

Glossary

acid precipitation
Rain or snow which contains dissolved gases from the air, forming dilute sulfuric and nitric acid.

adiabatic change
The warming of air when it is compressed, or cooling when it expands.

aerosols
Solid or liquid particles in the air; includes pollen, dust, chemicals, and sea salt.

air mass
A large volume of air with uniform temperature and humidity.

air pressure
Also called barometric pressure. A measure of the total weight of air above a certain area.

atmosphere
The envelope of air around the earth.

aurora
The northern and southern lights, caused by solar radiation hitting the earth's outer atmosphere.

climate
A long-time (30–100 years) average of weather conditions.

cloud
A visible cluster of liquid or frozen water droplets.

cold front
The boundary where cold air is pushing beneath warmer air.

condensation nuclei
Small particles of airborne dust, pollen, etc. around which water vapor condenses.

convection
Upward movement of air due to heating, expansion, and increased buoyancy.

Coriolis effect
The deflection of moving objects by the earth's rotation.

dew point
The temperature at which the air is saturated with moisture and produces dew.

evaporation
Conversion of water from the liquid to the gaseous state.

fossil fuel
Energy sources which include coal, oil, and gas.

front
The boundary between two air masses with different properties.

greenhouse effect
Atmospheric absorption and retention of solar radiation. As a result, the earth warms.

hail
Balls of ice with a layered, onion-like structure. They form in turbulent thunderclouds.

humidity
A measure of invisible water vaporized in air.

hydrologic cycle
The movement of water throughout the environment.

infrared radiation
A form of light with an energy slightly less than the visible colors, and a longer wavelength.

jet stream
A narrow stream of strong winds in the upper atmosphere.

lapse rate
The change of air temperature with height.

lightning
The rapid movement of electrical charge through the air.

meteorology
The study of the atmosphere, including weather prediction.

ozone layer
An upper atmospheric layer of triatomic oxygen, O_3.

relative humidity
The actual amount of water vapor in the air, compared with the maximum amount the air could hold if saturated. Half-saturated air has 50 percent relative humidity.

smog
Combination of smoke and fog. Sometimes ozone and carbon monoxide are also called smog.

spectrum
The total range of light, including all colors and energies.

temperature inversion
Cool air trapped beneath warmer air.

ultraviolet radiation
A form of light with an energy slightly greater than the visible colors, and a shorter wavelength.

warm front
The boundary where moving warm air is overriding cooler air.

water vapor
Atmospheric moisture in gaseous form.

weather
The state of the atmosphere at any given time.

wind
The movement of air.

wind chill
The loss of heat by exposed skin due to wind.

wind shear
Sudden changes in wind speed and direction.

Scripture Index

Note: References are to question numbers rather than page numbers.

Subject Index

Note: References are to question numbers rather than page numbers.